Solutions Manual

for

Exploring Chemical Analysis
Fifth Edition

Daniel C. Harris

W. H. Freeman and Company
New York

Solutions Manual to Accompany Harris Exploring Chemical Analysis, Fifth Edition

© 2012, 2009, 2005, 2001, 1997 by W. H. Freeman and Company

All rights reserved.

Printed in the United States of America

First Printing

Published under license, in the United States by
W. H. Freeman and Company
41 Madison Avenue
New York, NY 10010

www.whfreeman.com

ISBN-13: 978-1-4641-0641-5
ISBN-10: 1-4641-0641-X

Table of Contents

Chapter 0 – The Analytical Process 1

Chapter 1 – Chemical Measurements 2

Chapter 2 – Tools of the Trade 11

Chapter 3 – Math Toolkit 13

Chapter 4 – Statistics 21

Chapter 5 – Quality Assurance and Calibration Methods 30

Chapter 6 – Good Titrations 40

Chapter 7 – Gravimetric and Combustion Analysis 47

Chapter 8 – Introducing Acids and Bases 54

Chapter 9 – Buffers 62

Chapter 10 – Acid-Base Titrations 69

Chapter 11 – Polyprotic Acids and Bases 87

Chapter 12 – A Deeper Look at Chemical Equilibrium 104

Chapter 13 – EDTA Titrations 117

Chapter 14 – Electrode Potentials 129

Chapter 15 – Electrode Measurements 139

Chapter 16 – Redox Titrations 150

Chapter 17 – Instrumental Methods in Electrochemistry 161

Chapter 18 – Let There Be Light 171

Chapter 19 – Spectrophotometry: Instruments and Applications 181

Chapter 20 – Atomic Spectroscopy 190

Chapter 21 – Principles of Chromatography and Mass Spectrometry 199

Chapter 22 – Gas and Liquid Chromatography 209

Chapter 23 – Chromatographic Methods and Capillary Electrophoresis ... 222

CHAPTER 0
THE ANALYTICAL PROCESS

0-1. Qualitative analysis finds out *what* is in a sample. Quantitative analysis measures *how much* is in a sample.

0-2. Steps in a chemical analysis:
1. Formulate the question: Convert a general question into a specific one that can be answered by a chemical measurement.
2. Select the appropriate analytical procedure.
3. Obtain a representative sample.
4. Sample preparation: Convert the representative sample into a sample suitable for analysis. If necessary, concentrate the analyte and remove or mask interfering species.
5. Analysis: Measure the unknown concentration in replicate analyses.
6. Produce a clear report of results, including estimates of uncertainty.
7. Draw conclusions: Based on analytical results, decide what actions to take.

0-3. Masking converts an interfering species to a noninterfering species.

0-4. A calibration curve shows the response of an analytical method as a function of the known concentration of analyte in standard solutions. The calibration curve lets us deduce the concentration of an unknown from a measured response.

CHAPTER 1
CHEMICAL MEASUREMENTS

> *A note from Dan:* Don't worry if your numerical answers are slightly different from those in the *Solutions Manual*. You or I may have rounded intermediate results. In general, retain many extra digits for intermediate answers and save your roundoff until the end. We'll study this process in Chapter 3.

1-1. (a) meter (m), kilogram (kg), second (s), ampere (A), kelvin (K), mole (mol)

(b) hertz (Hz), newton (N), pascal (Pa), joule (J), watt (W)

1-2.
(a) mW = milliwatt = 10^{-3} watt
(b) pm = picometer = 10^{-12} meter
(c) kΩ = kiloohm = 10^{3} ohm
(d) μC = microcoulomb = 10^{-6} coulomb
(e) TJ = terajoule = 10^{12} joule
(f) ns = nanosecond = 10^{-9} second
(g) fg = femtogram = 10^{-15} gram
(h) dPa = decipascal = 10^{-1} pascal

1-3.
(a) 100 fJ or 0.1 pJ
(b) 43.172 8 nC
(c) 299.79 THz
(d) 0.1 nm or 100 pm
(e) 21 TW
(f) 0.483 amol or 483 zmol

1-4. (a) 1 horsepower = 745.700 W

100 horsepower = 100 × 745.700 W = 7.457×10^4 W

(b) 1 W = 1 J/s. Therefore, 7.457×10^4 W = 7.457×10^4 J/s

(c) With the conversion factor 1 cal/4.184 J, we can answer the question by using multiplication:

$$\left(7.457 \times 10^4 \,\frac{J}{s}\right)\left(\frac{1 \text{ cal}}{4.184 \text{ J}}\right) = 1.782 \times 10^4 \text{ cal/s}$$

If you write the conversion factor as 4.184 J/cal, you can divide instead of multiply to arrive at the same answer:

$$\frac{7.457 \times 10^4 \,\frac{J}{s}}{4.184 \,\frac{J}{\text{cal}}} = 1.782 \times 10^4 \text{ cal/s}$$

(d) $\left(1.782 \times 10^4 \dfrac{\text{cal}}{\text{s}}\right) \times \left(3\,600 \dfrac{\text{s}}{\text{h}}\right) = 6.416 \times 10^7 \text{ cal/h}$

1-5. (a) We know that there are exactly 0.025 4 m in 1 inch. We can set up a proportion relating the known quotient 0.025 4 m/inch to an unknown quotient:

$$\dfrac{0.025\,4 \text{ m}}{1 \text{ inch}} = \dfrac{1 \text{ m}}{x \text{ inch}} \Rightarrow x = 39.37 \text{ inches}$$

When you write a proportion, be sure the units are the same on both sides of the equation.

(b) $\left(345 \dfrac{\text{m}}{\text{s}}\right)\left(\dfrac{1 \text{ inch}}{0.025\,4 \text{ m}}\right)\left(\dfrac{1 \text{ foot}}{12 \text{ inch}}\right)\left(\dfrac{1 \text{ mile}}{5\,280 \text{ foot}}\right) = 0.214 \dfrac{\text{mile}}{\text{s}}$

$\left(0.214 \dfrac{\text{mile}}{\text{s}}\right)\left(3\,600 \dfrac{\text{s}}{\text{h}}\right) = 770 \dfrac{\text{mile}}{\text{h}}$

(c) $(3.00 \text{ s})\left(345 \dfrac{\text{m}}{\text{s}}\right) = 1.04 \times 10^3 \text{ m} = 1.04 \text{ km}$

$(1.04 \times 10^3 \text{ m})\left(\dfrac{1 \text{ inch}}{0.025\,4 \text{ m}}\right)\left(\dfrac{1 \text{ foot}}{12 \text{ inch}}\right)\left(\dfrac{1 \text{ mile}}{5\,280 \text{ foot}}\right) = 0.643 \text{ mile}$

1-6. (a) yN is yoctonewtons = 10^{-24} N; 174 yN = 174×10^{-24} N = 1.74×10^{-22} N

(b) 0.5 mK = 0.5×10^{-3} K = 5×10^{-4} K

(c) Atomic mass of Be = 9.012 g/mol

Mass of Be atom = $\dfrac{9.012 \text{ g/mol}}{6.022 \times 10^{23} \text{ mol}^{-1}} = 1.497 \times 10^{-23}$ g

Mass of 60 Be atoms = $(60)(1.497 \times 10^{-23} \text{ g}) = 8.979 \times 10^{-22}$ g

Mass of 60 Be atom in kg = $(8.979 \times 10^{-22} \text{ g})\left(\dfrac{1 \text{ kg}}{1\,000 \text{ g}}\right) = 8.979 \times 10^{-25}$ kg

Gravitational force = $\dfrac{GMm}{r^2}$

$= \dfrac{\left(6.674 \times 10^{-11} \dfrac{\text{m}^3}{\text{s}^2 \cdot \text{kg}}\right)(5.98 \times 10^{24} \text{ kg})(8.979 \times 10^{-25} \text{ kg})}{(6.38 \times 10^6 \text{ m})^2} = 8.804 \times 10^{-24} \dfrac{\text{kg} \cdot \text{m}}{\text{s}^2}$

In Table 1-2, we see that the units kg · m/s² are equivalent to newtons. The gravitational force is 8.804×10^{-24} N = 8.8 yN, which is about 5% of the electrical force in part (a).

1-7. (a) molarity = moles of solute / liter of solution
(b) molality = moles of solute / kilogram of solvent
(c) density = grams of substance / milliliter of substance

(d) weight percent = 100 × (mass of substance / mass of solution or mixture)

(e) volume percent = 100 × (volume of substance / volume of solution or mixture)

(f) parts per million = 10^6 × (mass of substance / mass of sample)

(g) parts per billion = 10^9 × (mass of substance / mass of sample)

(h) formal concentration = moles of formula / liter of solution

1-8. We need to find the moles of NaCl and divide by liters of solution.

32.0 g / [(22.989 770 + 35.452 7) g/mol] = 0.547 5 mol NaCl

0.547 5 mol / 0.500 L = 1.10 M

1-9. Mass of solution = (250 mL) (1.00 g/mL) = 250 g

$$\text{ppm} = \frac{13.7 \times 10^{-6} \text{ g}}{250 \text{ g}} \times 10^6 = 0.054\ 8 \text{ ppm}$$

$$\text{ppb} = \frac{13.7 \times 10^{-6} \text{ g}}{250 \text{ g}} \times 10^9 = 54.8 \text{ ppb}$$

1-10. $\dfrac{80 \times 10^{-3} \text{ g}}{180.2 \text{ g/mol}} = 4.4 \times 10^{-4}$ mol; $\dfrac{4.4 \times 10^{-4} \text{ mol}}{100 \times 10^{-3} \text{ L}} = 4.4 \times 10^{-3}$ M;

Similarly, 120 mg/L = 6.7×10^{-3} M

1-11. (a) The definition of weight percent is grams of solute per gram of total solution.

$$\left(0.705 \frac{\text{g HClO}_4}{\text{g solution}}\right)(100.0 \text{ g solution}) = 70.5 \text{ g HClO}_4$$

(b) There are 70.5 g of $HClO_4$ in 100.0 g of solution, so H_2O = 100.0 g − 70.5 g = 29.5 g

(c) mol $HClO_4$ = $\dfrac{70.5 \text{ g}}{100.458 \text{ g/mol}}$ = 0.702 mol

1-12. We can find the moles required to make 2.00 L of solution and then multiply moles by g/mol to get grams: $2.00 \text{ L} \times 0.050\ 0 \dfrac{\text{mol}}{\text{L}} \times 61.83 \dfrac{\text{g}}{\text{mol}} = 6.18 \text{ g}$

1-13. (a) From the definition of ppm, we can find grams of F^- per g of solution and then equate this as approximately equal to g F^-/mL of solution:

$$1.2 \text{ ppm } F^- = 1.2 \times 10^{-6} \frac{\text{g } F^-}{\text{g solution}} \approx 1.2 \times 10^{-6} \frac{\text{g } F^-}{\text{mL solution}}$$

The grams of F^- in 1 L are

Chemical Measurements

$$1.2 \times 10^{-6} \frac{\text{g F}^-}{\text{mL}} 1\,000 \frac{\text{mL}}{\text{L}} = 1.2 \times 10^{-3} \frac{\text{g F}^-}{\text{L}}$$

Next convert g/L to mol/L by dividing by formula mass:

$$\text{Molarity} = \left(1.2 \times 10^{-3} \frac{\text{g F}^-}{\text{L}}\right) / \left(18.998 \frac{\text{g F}^-}{\text{mol}}\right) = 6.32 \times 10^{-5} \text{ M}$$

We can use the dilution formula to find the required volume of reagent. But, we need to find the volume of the reservoir first.

Reservoir volume = $\pi r^2 h = \pi (50 \text{ m})^2 (20 \text{ m}) = 1.57 \times 10^5 \text{ m}^3$

But $1 \text{ m}^3 = 10^3 \text{ L} \Rightarrow (1.57 \times 10^5 \text{ m}^3)(10^3 \text{ L/m}^3) = 1.57 \times 10^8 \text{ L}$

Now use the dilution formula, remembering that 1.0 M H_2SiF_6 contains 6.0 M fluoride:

$M_{conc} \cdot V_{conc} = M_{dil} \cdot V_{dil}$

$(6.0 \text{ M})(x \text{ L}) = (6.32 \times 10^{-5} \text{ M})(1.57 \times 10^8 \text{ L}) \Rightarrow x = 1.7 \times 10^3 \text{ L}$

(b) mol H_2SiF_6 required = $\frac{1}{6}$ (mol F$^-$) because 1 mol H_2SiF_6 provides 6 mol F$^-$

mol $H_2SiF_6 = \frac{1}{6} (6.32 \times 10^{-5} \text{ mol/L})(1.57 \times 10^8 \text{ L}) = 1.65 \times 10^3 \text{ mol}$

g $H_2SiF_6 = (1.65 \times 10^3 \text{ mol})(144.09 \text{ g/mol}) = 2.4 \times 10^5 \text{ g}$

1-14. The meaning of 50 wt% NaOH is 0.50 g NaOH per gram of solution. To make use of this definition, we need to know how many grams of NaOH are required.

mol NaOH required = $1.00 \text{ L} \times 0.10 \frac{\text{mol}}{\text{L}} = 0.10 \text{ mol NaOH}$

g NaOH = 0.10 mol NaOH × 40.00 g/mol = 4.0 g NaOH

mass of solution containing 4.0 g NaOH = $\dfrac{4.0 \text{ g NaOH}}{0.50 \frac{\text{g NaOH}}{\text{g solution}}}$ = 8.0 g solution

1-15. (a) Use the dilution formula to find the volume of concentrated reagent required:

$$V_{conc} = V_{dil} \frac{M_{dil}}{M_{conc}} = 1\,000 \text{ mL} \left(\frac{1.00 \text{ M}}{18.0 \text{ M}}\right) = 55.6 \text{ mL}$$

(b) We want to find density, which is grams of solution per milliliter of solution. The meaning of 98.0 wt% H_2SO_4 is 0.980 g H_2SO_4 per gram of solution. One liter of 98.0 wt% H_2SO_4 contains $(18.0 \text{ mol})(98.079 \text{ g/mol}) = 1\,765 \text{ g}$ H_2SO_4. Because the solution contains 98.0 wt% H_2SO_4 and the mass of H_2SO_4 per mL is 1.765 g, the mass of solution per liter is

$$\frac{1.765 \text{ g } H_2SO_4/\text{mL}}{0.980 \text{ g } H_2SO_4/\text{g solution}} = 1.80 \text{ g solution/mL}$$

1-16. $\left(1.71 \dfrac{\text{mol CH}_3\text{OH}}{\text{L solution}}\right)(0.100 \text{ L solution}) = 0.171 \text{ mol CH}_3\text{OH}$

$(0.171 \text{ mol CH}_3\text{OH})\left(\dfrac{32.04 \text{ g}}{\text{mol CH}_3\text{OH}}\right) = 5.48 \text{ g}$

1-17. $1 \text{ ppm} = \dfrac{1 \text{ g solute}}{10^6 \text{ g solution}}$ Because 1 L of dilute solution $\approx 10^3$ g,

10^6 g solution $\approx 10^3$ L. Therefore, $1 \text{ ppm} = \dfrac{1 \text{ g solute}}{10^3 \text{ L}} = 10^{-3}$ g/L.

$1 \text{ ppm} = \left(10^{-3} \dfrac{\text{g}}{\text{L}}\right)\left(10^6 \dfrac{\mu\text{g}}{\text{g}}\right) = 10^3 \text{ }\mu\text{g/L}$

$1 \text{ ppm} = \left(10^3 \dfrac{\mu\text{g}}{\text{L}}\right)\left(10^{-3} \dfrac{\text{L}}{\text{mL}}\right) = 1 \text{ }\mu\text{g/mL}$

$1 \text{ ppm} = \left(10^3 \dfrac{\mu\text{g}}{\text{L}}\right)\left(10^{-3} \dfrac{\text{mg}}{\mu\text{g}}\right) = 1 \text{ mg/L}$

1-18. 0.2 ppb means 0.2×10^{-9} g of $C_{20}H_{42}$ per g of rainwater.

To convert to g of $C_{20}H_{42}$ per 1 000 g of rainwater, multiply by 1 000/1 000:

$\dfrac{0.2 \times 10^{-9} \text{ g C}_{20}\text{H}_{42}}{1 \text{ g rainwater}} \times \dfrac{1\ 000}{1\ 000} = \dfrac{0.2 \times 10^{-6} \text{ g C}_{20}\text{H}_{42}}{1\ 000 \text{ g rainwater}}$

With the approximation 1 g rainwater $\approx$ 1 mL rainwater, we can write

$\dfrac{0.2 \times 10^{-6} \text{ g C}_{20}\text{H}_{42}}{1\ 000 \text{ g rainwater}} \approx \dfrac{0.2 \times 10^{-6} \text{ g C}_{20}\text{H}_{42}}{\text{L rainwater}}$

To find molarity, divide g/L by molecular mass:

$\dfrac{0.2 \times 10^{-6} \text{ g/L}}{282.55 \text{ g/mol}} = 7 \times 10^{-10} \text{ M}$

1-19. We will convert ounces of tuna to grams of tuna and find out how many μg of mercury are in one can (6 oz) of tuna. From a body mass of 68 kg, we will compute how many days are allowed between eating tuna so that the average consumption does not exceed 0.1 μg Hg/kg body weight per day.

Table 1-4 tells us that 1 lb = 0.453 6 kg

6 oz = (6/16) lb = (6/16)(0.453 6 kg) = 0.170 kg = 170 g

One part per million means 1 μg Hg per gram of tuna. There is 0.6 ppm Hg in chunk white tuna = 0.6 μg Hg/g tuna.

A 6-oz can contains $(170 \text{ g tuna})\left(\dfrac{0.6 \text{ }\mu\text{g Hg}}{\text{g tuna}}\right) = 102 \text{ }\mu\text{g Hg}$

A dose of 0.1 μg Hg/kg body weight per day for a 68-kg person is

$$\left(\frac{0.1 \; \mu g \; Hg}{kg \cdot day}\right)(68 \; kg) = 6.8 \; \frac{\mu g \; Hg}{day}$$

If I eat 102 µg Hg in one day from one can of tuna, I have eaten the amount of mercury allowed in $(102 \; \mu g \; Hg)\left(\frac{1 \; day}{6.8 \; \mu g \; Hg}\right) = 15$ days

I should wait 15 days before consuming my next can of tuna so that my average intake does not exceed 6.8 µg Hg/day.

Chunk light tuna contains 0.14 ppm Hg = 0.14 µg Hg/g tuna. Substituting this number for 0.6 µg Hg/g tuna in the sequence of calculations gives a period of 3.5 days. I could eat 2 cans of chunk light tuna per week.

1-20. (a) Mass of solution $= 0.804 \; \frac{g}{mL} \times \frac{1 \; 000 \; mL}{L} = 804 \; \frac{g}{L}$

Mass of ethanol $= \frac{0.950 \; g \; of \; ethanol}{g \; of \; solution} \times \frac{804 \; g \; of \; solution}{L}$

$= 764 \; \frac{g \; of \; ethanol}{L}$

(b) $\dfrac{764 \; \frac{g}{L}}{46.07 \; \frac{g}{mol}} = 16.6 \; M$

1-21. (a) Remember that 10.2 wt% means (10.2 g solute)/(100 g solution), or (0.102 g solute)/(g solution). 10.0 g of 10.2 wt% solution contains

$0.102 \; \dfrac{g \; NiSO_4 \cdot 6H_2O}{g \; solution} \times 10.0 \; g \; solution = 1.02 \; g \; NiSO_4 \cdot 6H_2O$

$\dfrac{1.02 \; g \; NiSO_4 \cdot 6H_2O}{262.85 \; g \; NiSO_4 \cdot 6H_2O/mol} = 3.88 \times 10^{-3} \; mol \; NiSO_4 \cdot 6H_2O$

Now convert mol $NiSO_4 \cdot 6H_2O$ to g Ni:

$(3.88 \times 10^{-3} \; mol \; NiSO_4 \cdot 6H_2O)\left(1 \; \dfrac{mol \; Ni}{mol \; NiSO_4 \cdot 6H_2O}\right)\left(58.693 \; 4 \; \dfrac{g \; Ni}{mol \; Ni}\right)$

$= 0.228 \; g \; Ni$

(b) There are 0.412 mol of $NiSO_4 \cdot 6H_2O$ = 108.3 g of $NiSO_4 \cdot 6H_2O$ per L of solution. From the 10.2 wt%, we can say

$\dfrac{108.3 \; g \; NiSO_4 \cdot 6H_2O/L \; solution}{0.102 \; g \; NiSO_4 \cdot 6H_2O/g \; solution} = 1.06 \times 10^3 \; \dfrac{g \; solution}{L \; solution}$

Dividing numerator and denominator by 1 000 gives density $= 1.06 \; \dfrac{g}{mL}$

1-22. The mass of methanol in the solution is volume × density = (25.00 mL)(0.791 4 g/mL). The moles of methanol in the solution are mass/formula mass = (25.00 mL)(0.791 4 g/mL)/(32.042 g/mol). The molarity is (moles of methanol)/(total volume of solution). The final answer looks like this:

$$\frac{(25.00 \text{ mL})(0.791\ 4 \text{ g/mL}) / (32.042 \text{ g/mol})}{0.500\ 0 \text{ L}} = 1.235 \text{ M}$$

1-23. Find the volume of 12.1 M HCl reagent required by using the dilution formula:

$$M_{conc} \cdot V_{conc} = M_{dil} \cdot V_{dil}$$

$$12.1 \frac{\text{mol}}{\text{L}} \times V = 1.00 \frac{\text{mol}}{\text{L}} \times 0.100 \text{ L} \Rightarrow V = 8.26 \text{ mL}$$

Procedure: Dilute 8.26 mL of 12.1 M HCl to 100.0 mL in a volumetric flask.

1-24. (a) From the density of 1.43 g/mL, we can say 1 L = 1 430 g solution. In a 40.0 wt% solution, 40.0% of the mass is solute.

$0.400 \times 1\ 430 \text{ g} = 572 \text{ g CsCl}$ (572 g CsCl) / (168.36 g/mol) = 3.40 mol

Molar concentration = 3.40 mol / 1 L = 3.40 M

(b) $M_{conc} \cdot V_{conc} = M_{dil} \cdot V_{dil}$

(3.40 M) V = (0.100 M)(0.500 L) $\Rightarrow$ V = 14.7 mL

1-25. Shredded Wheat: 1 g contains 0.099 g protein + 0.799 g carbohydrate

$$0.099 \text{ g} \times 4.0 \frac{\text{Cal}}{\text{g}} + 0.799 \text{ g} \times 4.0 \frac{\text{Cal}}{\text{g}} = 3.6 \text{ Cal}$$

Doughnut: 1 g contains 0.046 g protein + 0.514 g carbohydrate + 0.186 g fat

$$0.046 \text{ g} \times 4.0 \frac{\text{Cal}}{\text{g}} + 0.514 \text{ g} \times 4.0 \frac{\text{Cal}}{\text{g}} + 0.186 \text{ g} \times 9.0 \frac{\text{Cal}}{\text{g}} = 3.9 \text{ Cal}$$

In a similar manner, we find 2.8 Cal/g for hamburger and 0.48 Cal/g for apple. Table 1-4 says that 16 ounces = 453.592 37 g $\Rightarrow$ 28.35 g/ounce. To convert Cal/g to Cal/ounce, multiply by 28.35:

	Shredded Wheat	Doughnut	Hamburger	Apple
Cal/g	3.6	3.9	2.8	0.48
Cal/ounce	102	111	79	14

1-26. Concentrations in equilibrium constants are expressed as dimensionless ratios of actual concentrations divided by standard-state concentrations, which are 1 M and 1 bar.

1-27. (a) $K = 1/[Ag^+]^3 [PO_4^{3-}]$ (b) $K = P_{CO_2}^6 / P_{O_2}^{15/2}$

1-28. (a) $P_A = \dfrac{2.8 \times 10^3 \text{ Pa}}{10^5 \text{ Pa/bar}} = 0.028$ bar

$P_E = \left(\dfrac{3.6 \times 10^4 \text{ torr}}{760 \text{ torr/atm}}\right)\left(\dfrac{101\,325 \text{ Pa/atm}}{10^5 \text{ Pa/bar}}\right) = 48$ bar

(b) $K = \dfrac{P_E^3}{P_A^2 [B]} = \dfrac{(48 \text{ bar})^3}{(0.028 \text{ bar})^2 (0.012 \text{ M})} = 1.2 \times 10^{10}$

1-29. It remains unchanged. $I_2(s)$ does not appear in the reaction quotient.

1-30.
$Cu^+ + N_3^- \rightleftharpoons CuN_3(s)$	$K_1 = 1/(4.9 \times 10^{-9})$
$HN_3 \rightleftharpoons H^+ + N_3^-$	$K_2 = 2.2 \times 10^{-5}$
$Cu^+ + HN_3 \rightleftharpoons CuN_3(s) + H^+$	$K_3 = K_1 K_2 = 4.5 \times 10^3$

1-31. (a)
$Ag^+ + Cl^- \rightleftharpoons AgCl(aq)$	$K_1 = 2.0 \times 10^3$
$AgCl(s) \rightleftharpoons Ag^+ + Cl^-$	$K_2 = 1.8 \times 10^{-10}$
$AgCl(s) \rightleftharpoons AgCl(aq)$	$K_3 = K_1 K_2 = 3.6 \times 10^{-7}$

(b) The answer to (a) tells us that $[AgCl(aq)] = 3.6 \times 10^{-7}$ M.

(c)
$AgCl_2^- \rightleftharpoons AgCl(aq) + Cl^-$	$K_1 = 1/93$
$Ag^+ + Cl^- \rightleftharpoons AgCl(s)$	$K_2 = 1/(1.8 \times 10^{-10})$
$AgCl(aq) \rightleftharpoons Ag^+ + Cl^-$	$K_3 = 1/(2.0 \times 10^3)$
$AgCl_2^- \rightleftharpoons AgCl(s) + Cl^-$	$K_4 = K_1 K_2 K_3 = 3.0 \times 10^4$

1-32. The concentration of zinc varies pretty smoothly with depth. One possible strategy to measure the average concentration is based on the approximation that the concentration of zinc at a depth of 50 m is representative of the concentration between 0 and 100 m. Then we could assume that the concentration at 150 m is representative of the concentration between 100 and 200 m, etc. To construct a representative sample of the 2000-m cylinder, we could take, say, 50 mL from a depth of 50 m, 50 mL from 150 m, 50 mL from 250 m, etc., down to 50 mL from a depth of 1 950 m. When the twenty 50-mL samples are combined, we have 1 L of liquid that is representative of the 2 000-m column of ocean. The concentration of zinc in the composite sample should be about the same as the average concentration in the 2 000-m cylinder of ocean water.

1-33. The mean volume of water flowing from the river in a year is

$$\left(\frac{560 \text{ cubic feet}}{\cancel{s}}\right)\left(\frac{3600 \cancel{s}}{\cancel{h}}\right)\left(\frac{24 \cancel{h}}{\cancel{day}}\right)\left(\frac{365 \cancel{day}}{yr}\right) = 1.766 \times 10^{10} \frac{\text{cubic feet}}{yr}$$

We need to convert cubic feet to L because we know how much nitrate is in each L. One foot is 12 inches = (12 in.)(2.54 cm/in.) = 30.48 cm. A cubic foot has a volume of 30.48 cm × 30.48 cm × 30.48 cm = 2.832×10^4 cm^3 = 2.832×10^4 mL = 28.32 L, so the mean volume per year is

$$\left(1.766 \times 10^{10} \frac{\cancel{ft^3}}{yr}\right)\left(28.32 \frac{L}{\cancel{ft^3}}\right) = 5.00 \times 10^{11} \frac{L}{yr}$$

The data given do not include the mean nitrate concentration. From the ranges 2.05–2.50 mg nitrate nitrogen/L in dry weather and 0.81–4.01 mg nitrate nitrogen/L in wet weather, I am going to assume that the mean concentration is 2.3 mg nitrate nitrogen/L. The formula mass of NO$_3^-$ is 62.00 g/mol, which contains one N atom (= 14.01 g/mol). For every 14.01 mg of nitrate nitrogen, there are 62.00 mg of nitrate anion. Therefore, a concentration of 2.3 mg nitrate nitrogen/L corresponds to

$$\left(\frac{62.00 \text{ mg nitrate}}{14.01 \cancel{\text{mg N}}}\right)\left(\frac{2.3 \cancel{\text{mg N}}}{L}\right) = 10.2 \frac{\text{mg NO}_3^-}{L}$$

The total flow of nitrate anion in a year is

$$\left(10.2 \frac{\text{mg NO}_3^-}{\cancel{L}}\right)\left(5.00 \times 10^{11} \frac{\cancel{L}}{yr}\right) = 5.10 \times 10^{12} \frac{\text{mg NO}_3^-}{yr}$$

Now we convert to metric tons:

$$\left(5.10 \times 10^{12} \frac{\cancel{mg}}{yr}\right)\left(\frac{1 \cancel{g}}{1\,000 \cancel{mg}}\right)\left(\frac{1 \text{ ton}}{1\,000 \times 10^3 \cancel{g}}\right) = 5.10 \times 10^3 \text{ tons/yr}$$

Because the mean concentration of nitrate in river water was just an estimate from the data given in the report, I would round the answer to ~5 000 tons/yr.

CHAPTER 2
TOOLS OF THE TRADE

2-1. The internal calibration mass enables the balance to know how much electric current in the electromagnetic coil is required to balance a known mass. This conversion factor is then applied to subsequent weighings of unknown masses.

2-2. TD means "to deliver" and TC means "to contain."

2-3. The plastic flask is needed for trace analysis on analytes at ppb levels that might be lost by adsorption on the glass surface. A plastic flask is also required for reagents such as HF or hot, basic solutions that react with glass.

2-4. The trap prevents backup of filtrate into the vacuum system or backup of water from the aspirator into the suction flask. The watchglass keeps dust out of the sample.

2-5. An **ab**sorbed substance is taken inside a material. An **ad**sorbed substance is bound to the surface. In drying glassware, we remove adsorbed water.

2-6. Digestion is the process in which a substance is dissolved by decomposition to smaller molecules or ions. In extraction, analyte is dissolved in solvent that does not necessarily dissolve the entire sample and does not decompose the analyte.

2-7. We use the buoyancy equation with the density of air $d_a = 0.001\ 2$ g/mL, the density of standard weights, $d_w = 8.0$ g/mL, the density of water, $d = 1.00$ g/mL, and the mass of water measured in air is $m' = 5.397\ 4$ g. The true mass that would be measured in vacuum is

$$m = \frac{(5.397\ 4\ \text{g})\left(1 - \dfrac{0.001\ 2\ \text{g/mL}}{8.0\ \text{g/mL}}\right)}{\left(1 - \dfrac{0.001\ 2\ \text{g/mL}}{1.00\ \text{g/mL}}\right)} = 5.403\ 1\ \text{g}$$

2-8. $$m = \frac{(14.82\ \text{g})\left(1 - \dfrac{0.001\ 2\ \text{g/mL}}{8.0\ \text{g/mL}}\right)}{\left(1 - \dfrac{0.001\ 2\ \text{g/mL}}{0.626\ \text{g/mL}}\right)} = 14.85\ \text{g}$$

2-9. $$m = \frac{(0.296\ 1\ \text{g})\left(1 - \dfrac{0.001\ 2\ \text{g/mL}}{8.0\ \text{g/mL}}\right)}{\left(1 - \dfrac{0.001\ 2\ \text{g/mL}}{5.24\ \text{g/mL}}\right)} = 0.296\ 1\ \text{g}$$

2-10. The mass delivered (measured in air) is 20.214 4 g − 10.263 4 g = 9.951 0 g. Table 2-5 gives the correction factor (1.002 9 mL/g at 20°C) to change measured mass into true volume.

True volume = (9.951 0 g)(1.002 9 mL/g) = 9.980 mL

2-11. Correction factor in Table 2-5 at 26°C is 1.004 3 mL/g.

Measured mass delivered = 14.974 g − 9.974 g = 5.000 g

True volume = (5.000 g)(1.004 3 mL/g) = 5.022 mL

2-12. Correction factor in Table 2-5 at 25°C is 1.004 0 mL/g.

True volume = (15.569 g)(1.004 0 mL/g) = 15.631 mL

2-13. Our strategy is to find the total aluminum content of the glass from the experiment in which the glass was completely dissolved and analyzed. We compare that to the amount of aluminum extracted from the glass by EDTA.

We know that the total Al content of the glass is 0.80 wt%, which means 0.008 0 g Al per gram of glass. In 0.5 g of glass, there would be

(0.008 0 g Al/g glass)(0.50 g glass) = 0.004 0 g Al

200 mL of EDTA solution extracted enough Al from 0.50 g of glass to give a final concentration of 5.2 μM Al after two months.

mol Al extracted = (0.20 L)(5.2 × 10^{-6} mol/L) = 1.04 × 10^{-6} mol

mass of Al extracted = (1.04 × 10^{-6} mol)(27.0 g/mol) = 2.81 × 10^{-5} g

fraction of Al extracted = $\dfrac{\text{mass of Al removed by EDTA}}{\text{total mass of Al present in glass}}$

= $\dfrac{2.81 \times 10^{-5} \text{ g}}{0.004 \, 0 \text{ g}}$ = 0.007 0 = 0.70%

CHAPTER 3
MATH TOOLKIT

3-1. (a) 1.237 (b) 1.238 (c) 0.135 (d) 2.1 (e) 2.00

3-2. (a) 0.217 (b) 0.216 (c) 0.217 (d) 0.216

3-3. (a) 4 (b) 4 (c) 4

3-4. (a) 12.3 (b) 75.5 (c) 5.520×10^3 (d) 3.04
(e) 3.04×10^{-10} (f) 11.9 (g) 4.600 (h) 4.9×10^{-7}

3-5. (a) 12.01 (b) 10.9 (c) 14 (d) 14.3
(e) −17.66 (f) 5.97×10^{-3} (g) 2.79×10^{-5}

3-6. (a) $137.327 \times 1 = 137.327$ Ba (b) $12.0107 \times 31 = 372.331_7$ C_{31}
$35.453 \times 2 = \underline{70.906}$ Cl_2 $1.00794 \times 32 = 32.2541$ H_{32}
208.233 $BaCl_2$ $15.9994 \times 8 = 127.995_2$ O_8
$14.0067 \times 2 = \underline{28.0134}$ N_2
560.594 $C_{31}H_{32}O_8N_2$

Multiply each atomic mass by the number of atoms in the formula. For example, in $C_{31}H_{32}O_8N_2$, multiply the atomic mass of carbon by 31. Because the atomic mass of carbon (12.010 7) has 6 significant digits, the product 12.0107×31 is limited to 6 digits and we wrote the next digit as a subscript (372.331_7). When adding the masses, terminate the sum at the decimal place of the last significant digit in the term with the fewest decimal places.

3-7. $2 \times Mn = 2 \times 54.938\,045 = 109.876\,090$
$10 \times C = 10 \times 12.0107 = 120.107$ ← 6 significant figures based on 12.010 7
$10 \times O = 10 \times 15.9994 = \underline{159.994}$ ← 6 significant figures based on 15.999 4
$Mn_2(CO)_{10}$ 389.977 ← limited to 3rd decimal place by C and O

3-8. Because all measurements have some uncertainty, there is no way to know the true value.

3-9. (a) Systematic error arises from a procedure or an instrument that gives inaccurate results. The error is always in the same direction (positive or negative) if you repeat the same measurement. With care, systematic error can be detected and corrected. Random error arises from physical limitations in making measurements. It is sometimes negative and sometimes positive when repeating the same measurement.

(b) 25.031 mL is a systematic error. The pipet always delivers more than it is rated for. The number ± 0.009 is the random error in the volume delivered. The volume fluctuates around 25.031 by ± 0.009 mL.

(c) The numbers 1.98 and 2.03 mL are systematic errors. The buret delivers too little between 0 and 2 mL and too much between 2 and 4 mL. The observed variations ± 0.01 and ± 0.02 are random errors.

(d) The difference between 1.9839 and 1.9900 g is random error. The mass will probably be different the next time I try the same procedure.

(e) Differences in peak area are random error based on inconsistent injection volume, inconsistent detector response, and probably other small variations in the condition of the instrument from run to run.

(f) The funnel was not heated before the first use, so it probably had adsorbed moisture. The calculated mass of precipitate is probably low because the empty funnel would have weighed less if I had dried it before weighing. This is systematic error in my procedure. There is also random error in every measurement. We cannot determine it from only one experiment.

3-10. $0.167\,89\%$ of $3.123\,56 = 0.005\,24$ The uncertainty lies in the third decimal place, so the answer can be written (a) 3.124 ± 0.005; (b) $3.124 \pm 0.2\%$. It would also be reasonable to keep an additional digit: (a) $3.123_6 \pm 0.005_2$; (b) $3.123_6 \pm 0.1_7\%$.

3-11. (a) $\quad 6.2\ (\pm 0.2)$
$\quad\quad\quad -\underline{4.1\ (\pm 0.1)}$
$\quad\quad\quad 2.1 \pm e^2 = 0.2^2 + 0.1^2 \Rightarrow e = 0.2_2$

Answer: 2.1 ± 0.2 (or $2.1 \pm 11\%$) (If you did not keep all the digits in 0.2_{23606}, uncertainty would be 10% instead of 11%. Either answer is fine.)

(b) $\quad 9.43\ (\pm 0.05) \quad\quad\quad 9.43\ (\pm 0.53\%)$
$\quad\quad \times\ \underline{0.016\ (\pm 0.001)} \Rightarrow \times\ \underline{0.016\ (\pm 6.25\%)} \quad\quad \%e^2 = 0.53^2 + 6.25^2$
$\quad\quad\quad\quad\quad\quad\quad\quad\quad\quad\quad 0.150\,88\ (\pm\%e) \quad\Rightarrow \%e = 6.27_2$

Relative uncertainty = 6.27%
Absolute uncertainty = $0.150\,88 \times 0.062\,7 = 0.009\,46$
Answer: 0.151 ± 0.009 (or $0.151 \pm 6\%$)

(c) The first term in brackets is the same as part (a), so we can rewrite the problem as 2.1 ($\pm 0.2_{24}$) ÷ 9.43 (± 0.05) = 2.1 ($\pm 10.7\%$) ÷ 9.43 ($\pm 0.53\%$).
$\%e = \sqrt{10.7^2 + 0.53^2} = 10.7\%$
Absolute uncertainty = $0.107 \times 0.223 = 0.023\,9$
Answer: $0.22_3 \pm 0.02_4$ ($\pm 11\%$)

(d) The term in braces is
$\quad 6.2\ (\pm 0.2) \times 10^{-3}$
$\underline{+ 4.1\ (\pm 0.1) \times 10^{-3}}$
$10.3\ (\pm 0.2_{24}) \times 10^{-3} = 10.3 \times 10^{-3}\ (\pm 2.2\%)$

$e = \sqrt{0.2^2 + 0.1^2} \Rightarrow e = 0.224$

$9.43\ (\pm 0.53\%) \times 0.0103\ (\pm 2.2\%) = 0.097\,13 \pm 2.26\% = 0.097\,13 \pm 0.002\,20$
Answer: $0.097_1 \pm 0.002_2$ ($\pm 2._3\%$)

3-12. (a) Use relative uncertainties for multiplication and division:
$[12.41\ (\pm 0.09) \div 4.16\ (\pm 0.01)] \times 7.068\,2\ (\pm 0.000\,4)$
$= [12.41\ (\pm 0.73\%) \div 4.16\ (\pm 0.24\%)] \times 7.068\,2\ (\pm 0.005\,7\%)$
$= 21.09 \pm 0.77\% = 21.09 \pm 0.16$
(because $\sqrt{0.73\%^2 + 0.24\%^2 + 0.005\,7\%^2} = 0.77\%$)
Answer: $21.0_9 \pm 0.1_6$; relative uncertainty = $0.7_7\%$
or 21.1 ± 0.2; relative uncertainty = 0.8%

(b) First use relative uncertainty for the product in brackets:
$[3.26\ (\pm 0.10) \times 8.47\ (\pm 0.05)] = [3.26\ (\pm 3.07\%) \times 8.47\ (\pm 0.59\%)]$
$= 27.61 \pm 3.13\%$ (because $\sqrt{3.07\%^2 + 0.59\%^2} = 3.13\%$)
$= 27.61 \pm 0.86$
Then use absolute uncertainties for the subtraction:
$[27.61 \pm 0.86] - 0.18\ (\pm 0.06) = 27.43 \pm 0.86$
A good answer is $27.4_3 \pm 0.8_6$; relative uncertainty = $0.8_6 / 27.4_3 = 3._1\%$

(c) First use absolute uncertainty for the subtraction in brackets:
$[2.09\ (\pm 0.04) - 1.63\ (\pm 0.01)] = 0.46\ (\pm 0.041)$
(because $\sqrt{0.04^2 + 0.01^2} = 0.041$)
Then use relative uncertainties for the division:
$6.843\ (\pm 0.008) \times 10^4 \div [0.46\ (\pm 0.041)]$
$= 6.843\ (\pm 0.12\%) \times 10^4 \div [0.46\ (\pm 8.9\%)] = (14.87 \pm 8.9\%) \times 10^4$
$= (14._9 \pm 1._3) \times 10^4$ or $(15 \pm 1) \times 10^4$; relative uncertainty = 9%

3-13. (a) $\sqrt{0.03^2 + 0.02^2 + 0.06^2} = 0.07 \Rightarrow$ Answer: 10.18 (±0.07) (±0.7%)

(b) 91.3 (±1.10%) × 40.3 (±0.50%) / 21.2 (±0.94%)

$\sqrt{1.10^2 + 0.50^2 + 0.94^2} = 1.53 \Rightarrow$ Answer: 174 (±3) (±2%)

(c) 4.97 (±0.05) − 1.86 (±0.01) = 3.11 (±0.051)

3.11 (±1.64%) / 21.2 ± (0.94%) = 0.147 ± 1.9%

Answer: 0.147 (±0.003) (±2%)

(d)
```
   2.0164    (±0.0008)
   1.233     (±0.002)
 + 4.61      (±0.01)
 ─────────
   7.85₉₄   (± √(0.0008)² + (0.002)² + (0.01)² = 0.01₀₂)
```

Answer: 7.86 (±0.01) (±0.1%)

(e)
```
   2016.4   (±0.8)
  + 123.3   (±0.2)
  +  46.1   (±0.1)
  ─────────
   2185.8   (± √(0.8)²+(0.2)²+(0.1)² = 0.8)
```

Answer: 2185.8 (±0.8) (±0.04%)

3-14. (a) 3.4 ± 0.2 $e = \sqrt{0.2^2 + 0.1^2} = 0.224$
+ 2.6 ± 0.1

6.0 ± e = 6.0 ± 0.2₂₄ (±3.7%) or 6.0 ± 0.2 (±4%)

It is all right to leave extra digits if you subscript them to show that you know they are not significant. We might keep extra digits in case we may need to use the number for further calculations.

(b) $\dfrac{3.4 \pm 0.2}{2.6 \pm 0.1} = \dfrac{3.4 \pm 5.88\%}{2.6 \pm 3.85\%} = 1.308 \pm e$

%$e = \sqrt{5.88^2 + 3.85^2} = 7.03\%$

Answer: 1.30₈ ± 0.09₂ (±7.0%)

(c) $\dfrac{3.4\ (\pm 0.2) \times 10^{-8}}{2.6\ (\pm 0.1) \times 10^{3}} = \dfrac{3.4\ (\pm 5.88\%) \times 10^{-8}}{2.6\ (\pm 3.85\%) \times 10^{3}}$

= 1.30₈ (±0.09₂) × 10⁻¹¹ (±7.0%)

(d) 3.4 (±0.2) − 2.6 (±0.1) = 0.8 ± 0.2₂₄ = 0.8 ± 28.0%

0.8 (±28.0%) × 3.4 (±5.88%) = 2.72 ± 28.6%

Answer: 2.7₂ ± 0.7₈ (±29%)

3-15. (a) For pH = 5.42 ± 0.05, $[H^+] = 10^{-5.42} = 3.802 \times 10^{-6}$ M
Uncertainty in $[H^+]$ = 2.303$[H^+]$(uncertainty in pH)
$= 2.303[3.802 \times 10^{-6}$ M$](0.05) = 4.38 \times 10^{-7}$ M $= 0.438 \times 10^{-6}$ M
$[H^+] = 3.8\ (\pm 0.4) \times 10^{-6}$ M

(b) Relative uncertainty in $[H^+] = (0.438 \times 10^{-6}$ M$)/(3.802 \times 10^{-6}$ M$) = 12\%$

(c) Uncertainty in $[H^+]$ = 2.303$[H^+]$(uncertainty in pH)
Relative uncertainty $= \dfrac{\text{uncertainty in }[H^+]}{[H^+]} = 2.303$(uncertainty in pH)

That is, the relative uncertainty is proportional to the uncertainty in pH, but does not depend on $[H^+]$, and therefore does not depend on pH.

3-16. For pH = 8.2 ± 0.1, $[H^+] = 10^{-8.2} = 6.310 \times 10^{-9}$ M
Absolute uncertainty in $[H^+]$ = 2.303$[H^+]$(uncertainty in pH)
$= 2.303[6.31 \times 10^{-9}$ M$](0.1) = 1.453 \times 10^{-9}$ M
$[H^+] = 6.3\ (\pm 1.5) \times 10^{-9}$ M (or $6\ (\pm 1) \times 10^{-9}$ M)
Relative uncertainty = $(1.453 \times 10^{-9}$ M$)/(6.310 \times 10^{-9}$ M$) = 23\%$

3-17. 6C: $6 \times 12.0107 \pm 0.0008 = 72.0642 \pm 0.0048$
6H: $6 \times 1.00794 \pm 0.00007 = 6.04764 \pm 0.00042$
uncertainty = $\sqrt{0.0048^2 + 0.00042^2} = 0.005$
FM = 78.112 ± 0.005

3-18.
$6C = 6 \times 12.0107\ (\pm 0.0008) = 72.0642\ (\pm 0.0048)$
$13H = 13 \times 1.00794\ (\pm 0.00007) = 13.10322\ (\pm 0.00091)$
$B = \underline{\qquad\qquad\qquad\qquad\quad 10.811\qquad (\pm 0.007)\qquad}$
$\qquad\qquad\qquad\qquad\qquad\quad 95.978 \pm \sqrt{0.0048^2 + 0.00091^2 + 0.007^2}$

Answer: FM = 95.978 ± 0.009

3-19. (a) Na = $\quad 22.989770 \quad \pm 0.000002$ g/mol
$\qquad$ Cl = $\quad \underline{35.453 \qquad\quad \pm 0.002 \quad\text{g/mol}}$
$\qquad\qquad\quad 58.442770 \quad \sqrt{(2 \times 10^{-6})^2 + (2 \times 10^{-3})^2} = 2 \times 10^{-3}$
FM = 58.443 ± 0.002 g/mol

(b) Molarity $= \dfrac{\text{mol}}{\text{L}} = \dfrac{[2.634\ (\pm 0.002)\text{g}]\ /\ [58.443\ (\pm 0.002)\text{ g/mol}]}{0.10000\ (\pm 0.00008)\text{ L}}$

$= \dfrac{[2.634\ (\pm 0.076\%)]\ /\ [58.4425\ (\pm 0.0034\%)]}{0.10000\ (\pm 0.08\%)}$

Relative error = $\sqrt{(0.076\%)^2 + (0.0034\%)^2 + (0.08\%)^2} = 0.11\%$

Molarity = 0.4507 ± 0.0005 M

3-20. (a) Molarity = $\dfrac{0.2222\ (\pm 0.090\%)\ \text{g}}{214.0010\ (\pm 0.00042\%)\ \dfrac{\text{g}}{\text{mol}} \times 0.05000\ (\pm 0.10\%)\ \text{L}}$

$\%e = \sqrt{0.090^2 + 0.00042^2 + 0.10^2} = 0.135\%$

Molarity = 0.020766 ± 0.000028 M (or 0.02077 ± 0.00003 M)

(b) The uncertainty in the analysis is ~0.1%, so 0.1% uncertainty in reagent purity is significant.

3-21. Let's first consider how to work the problem without uncertanties. The mass of methanol in the solution is volume × density = (25.00 mL)(0.7914 g/mL). The moles of methanol in the solution are mass/formula mass = (25.00 mL)(0.7914 g/mL)/(32.0419 g/mol). The molarity is (moles of methanol)/(total volume of solution). The final arithmetic looks like this:

$\dfrac{(25.00\ \text{mL})(0.7914\ \text{g/mL}) / (32.0419\ \text{g/mol})}{0.5000\ \text{L}} = 1.235$ M

Now rewrite the same arithmetic with uncertainties:

$\dfrac{(25.00 \pm 0.03\ \text{mL})(0.7914 \pm 0.0002\ \text{g/mL}) / (32.0419 \pm 0.0009\ \text{g/mol})}{0.5000 \pm 0.0002\ \text{L}}$

For multiplication and division, convert absolute uncertainties to relative uncertainties:

$\dfrac{(25.00 \pm 0.12\%\ \text{mL})(0.7914 \pm 0.025\%\ \text{g/mL}) / (32.0419 \pm 0.003\%\ \text{g/mol})}{0.5000 \pm 0.040\ \text{L}}$

Relative uncertainty = $\sqrt{0.12^2 + 0.025^2 + 0.003^2 + 0.040^2} = 0.129\%$

Absolute uncertainty = 0.129% of 1.235 M = 0.002 M

Answer: 1.235 ± 0.002 M

3-22. (a) 2.00 L of 0.169 M NaOH (FM = 39.9971) requires 0.338 mol = 13.52 g NaOH.

$$\frac{13.52 \text{ g NaOH}}{0.534 \text{ g NaOH/g solution}} = 25.32 \text{ g solution}$$

$$\frac{25.32 \text{ g solution}}{1.52 \text{ g solution/mL solution}} = 16.6_6 \text{ mL}$$

(b) Molarity =

$$\frac{(16.6_6 \, (\pm 0.10) \text{ mL}) \left[1.52 \, (\pm 0.01) \frac{\text{g soln}}{\text{mL}} \right] \left[0.534 \, (\pm 0.004) \frac{\text{g NaOH}}{\text{g soln}} \right]}{\left[39.9971 \frac{\text{g NaOH}}{\text{mol}} \right] (2.00 \text{ L})}$$

Because the relative errors in formula mass and final volume are negligible (≈ 0), we can write

$$\text{Relative error in molarity} = \sqrt{\left(\frac{0.10}{16.66}\right)^2 + \left(\frac{0.01}{1.52}\right)^2 + \left(\frac{0.004}{0.534}\right)^2} = 1.16\%$$

Molarity = 0.169 ($\pm$0.002) M

3-23. Formula in cell F3: =B3*$A$4+C3*$A$6+D3*$A$8+E3*$A$10

3-24. Your graph should look like the one in the book.

3-25. We need to find the uncertainty in molarity of $AgNO_3$ prepared by each method. The uncertainty in mass is ± 0.0003 g. The uncertainty in a 100-mL Class A volumetric flask is ± 0.08 mL. The uncertainty in a 10-mL Class A transfer pipet is ± 0.02 mL.

The formula mass of $AgNO_3$ is

Ag:	107.8682 ± 0.0002
N:	14.0067 ± 0.0002
3 × O:	$3 \times (15.9994 \pm 0.0003) = 47.9982 \pm 0.0009$
$AgNO_3$:	$169.8731 \pm 0.0009_4$ g/mol

In Method 1, the molarity of silver nitrate is

$$\frac{\text{moles}}{\text{liters}} = \frac{(0.046\,3 \pm 0.000\,3 \text{ g})/(169.873\,1 \pm 0.000\,9_4 \text{ g/mol})}{0.100\,00 \pm 0.000\,08 \text{ L}}$$

$$= \frac{(0.046\,3 \pm 0.648\%)/(169.873\,1 \pm 0.000\,56\%)}{0.100\,00 \pm 0.08\%}$$

$$= 0.002\,726 \pm 0.653\% = 0.002\,72_6 \pm 0.000\,01_8 \text{ M}$$

In Method 2, the molarity of silver nitrate in the first flask is

$$\frac{\text{moles}}{\text{liters}} = \frac{(0.463\,0 \pm 0.000\,3 \text{ g})/(169.873\,1 \pm 0.000\,9_4 \text{ g/mol})}{0.100\,00 \pm 0.000\,08 \text{ L}}$$

$$= \frac{(0.463\,0 \pm 0.064\,8\%)/(169.873\,1 \pm 0.000\,54\%)}{0.100\,00 \pm 0.08\%}$$

$$= 0.027\,26\,(\pm 0.1_{03}\%) \text{ M} = 0.027\,26 \pm 0.000\,02_8 \text{ M}$$

The concentration in the second flask is

$$\frac{10.00 \pm 0.02 \text{ mL}}{100.00 \pm 0.08 \text{ mL}} \times (0.027\,26 \pm 0.000\,02_8 \text{ M})$$

$$= 0.002\,726 \pm 0.238\% = 0.002\,726 \pm 0.000\,006_5 \text{ M}$$

Method 2 is more accurate by about a factor of 3. The biggest difference is the large relative uncertainty in mass of AgNO$_3$ measured out in Method 1. This uncertainty of 0.6$_{48}$% is far larger than any of the other uncertainties in any quantity in either method.

CHAPTER 4
STATISTICS

4-1. The smaller the standard deviation, the greater the precision. There is no necessary relationship between standard deviation and accuracy. The statistics that we do in this chapter pertain to precision, not accuracy.

4-2. 0.683, 0.955, 0.997

4-3. Compute F and compare it with the value in the table.

$F_{\text{calculated}} = \dfrac{s_1^2}{s_2^2} = \dfrac{(0.47)^2}{(0.28)^2} = 2.8_2$. In Table 4-3, look in the column for $n - 1 = 10 - 1 = 9$ degrees of freedom and in the row for 9 degrees of freedom to find $F_{\text{table}} = 3.18$. Since $F_{\text{calculated}}\ (= 2.8_2) < F_{\text{table}}$, the difference is <u>not</u> significant at the 95% confidence level.

4-4. (a) Mean $= \dfrac{1}{8}(1.526\,60 + 1.529\,74 + 1.525\,92 + 1.527\,31 + 1.528\,94 +$
$1.528\,04 + 1.526\,85 + 1.527\,93) = 1.527\,67$

(b) Standard deviation =

$\sqrt{\dfrac{(1.526\,60 - 1.527\,67)^2 + \cdots + (1.527\,93 - 1.527\,67)^2}{8 - 1}} = 0.001\,26$

Relative standard deviation $= (0.001\,26\,/\,1.527\,67) \times 100 = 0.082\,5\%$

(c) $\mu = \bar{x} \pm \dfrac{ts}{\sqrt{n}} = 1.527\,93 \pm \dfrac{(3.707)(0.000\,07)}{\sqrt{7}} = 1.527\,93 \pm 0.000\,10$

4-5. (a) The confidence interval is a region around the measured mean in which the true mean is likely to lie.

(b) A 95% confidence interval is necessarily larger than a 90% confidence interval. The only way to increase the probability of finding the true mean in the interval is to increase the size of the interval.

4-6. Mean $= \dfrac{1}{5}(116.0 + 97.9 + 114.2 + 106.8 + 108.3) = 108.6_4$

Standard deviation $= \sqrt{\dfrac{(116.0 - 108.6_4)^2 + \cdots + (783 - 108.6_4)^2}{5 - 1}}$

$= 7.1_4$

90% confidence interval $= 108.6_4 \pm \dfrac{(2.132)(7.1_4)}{\sqrt{5}} = 108.6_4 \pm 6.8_1$

21

4-7. (a) $\bar{x}_1 = 0.02756$, $s_1 = 0.000488$; $\bar{x}_2 = 0.02690$, $s_2 = 0.000406$

(b) $F_{calculated} = \dfrac{s_1^2}{s_2^2} = \dfrac{(0.000488)^2}{(0.000406)^2} = 1.44$

For 4 degrees of freedom in the numerator and denominator, $F_{table} = 6.39$. $F_{calculated} < F_{table}$, so the standard deviations are <u>not</u> significantly different.

(c) Because the standard deviations are not significantly different, we use Equations 4-6 and 4-5 for the t test.

$$s_{pooled} = \sqrt{\dfrac{s_1^2(n_1 - 1) + s_2^2(n_2 - 1)}{n_1 + n_2 - 2}}$$

$$= \sqrt{\dfrac{0.000488^2(5-1) + 0.000406^2(5-1)}{5+5-2}} = 0.000449$$

$$t = \dfrac{|0.02756 - 0.02690|}{0.000449}\sqrt{\dfrac{5 \cdot 5}{5+5}} = 2.32$$

Because $t_{calculated} = 2.32 > t_{table}(95\%) = 2.306$, the difference <u>is</u> significant.

4-8. $\bar{x} = 2.29947$ g, $s = 0.00138$ g, 7 degrees of freedom

95% confidence: $\mu = \bar{x} \pm \dfrac{(2.365)(0.00138)}{\sqrt{8}} = 2.29947 \pm 0.00115$

99% confidence: $\mu = \bar{x} \pm \dfrac{(3.500)(0.00138)}{\sqrt{8}} = 2.29947 \pm 0.00171$

4-9. $\bar{x}_1 = 147.8$, $s_1 = 11.12$, $\bar{x}_2 = 157.2$, $s_2 = 5.89$

$$F_{calculated} = \dfrac{s_1^2}{s_2^2} = \dfrac{(11.12)^2}{(5.89)^2} = 3.56$$

For 4 degrees of freedom in the numerator and denominator, $F_{table} = 6.39$. $F_{calculated} < F_{table}$, so the standard deviations are <u>not</u> significantly different.

$$s_{pooled} = \sqrt{\dfrac{11.12^2 \cdot 4 + 5.89^2 \cdot 4}{5+5-2}} = 8.90$$

$$t = \dfrac{|157.2 - 147.8|}{8.90}\sqrt{\dfrac{5 \cdot 5}{5+5}} = 1.67 < 2.306 \text{ (Student's } t \text{ for 8 degrees of freedom)}$$

The difference is <u>not</u> significant.

4-10. (a) For indicators 1 and 2:

$$F_{\text{calculated}} = \frac{s_1^2}{s_2^2} = \frac{0.002\ 25^2}{0.000\ 98^2} = 5.2_7$$

For 27 degrees of freedom in the numerator and 17 in the denominator, $F_{\text{table}} \approx 2.04$, which we find under 30 degrees of freedom and across from 20 degrees of freedom. (The value found with the Excel function FINV(0.05,27,17) is 2.17.) $F_{\text{calculated}} > F_{\text{table}}$, so the standard deviations <u>are</u> significantly different.

(b) $$t_{\text{calculated}} = \frac{|\bar{x}_1 - \bar{x}_2|}{\sqrt{s_1^2/n_1 + s_2^2/n_2}} = \frac{|0.095\ 65 - 0.086\ 86|}{\sqrt{0.002\ 25^2/28 + 0.000\ 98^2/18}} = 18.2$$

$$\text{Degrees of freedom} = \frac{(s_1^2/n_1 + s_2^2/n_2)^2}{\dfrac{(s_1^2/n_1)^2}{n_1 - 1} + \dfrac{(s_2^2/n_2)^2}{n_2 - 1}}$$

$$= \frac{(0.002\ 25^2/28 + 0.000\ 98^2/18)^2}{\dfrac{(0.002\ 25^2/28)^2}{28 - 1} + \dfrac{(0.000\ 98^2/18)^2}{18 - 1}} = 39.8 \approx 40$$

For 40 degrees of freedom, the critical value of t in Table 4-4 for 95% confidence is 2.021. The observed value $t_{\text{calculated}} = 18.2$ far exceeds t_{table}, and is highly significant.

(c) For indicators 2 and 3:

$$F_{\text{calculated}} = \frac{s_1^2}{s_2^2} = \frac{0.001\ 13^2}{0.000\ 98^2} = 1.3_3$$

For 28 degrees of freedom in the numerator and 17 in the denominator, $F_{\text{table}} \approx 2.04$, which we find under 30 degrees of freedom and across from 20 degrees of freedom. (The value found with FINV(0.05,28,17) is 2.16.) $F_{\text{calculated}} > F_{\text{table}}$, so the standard deviations are <u>not</u> significantly different.

(d) $$s_{\text{pooled}} = \sqrt{\frac{0.001\ 13^2 \cdot 28 + 0.000\ 98^2 \cdot 17}{29 + 18 - 2}} = 0.001\ 08$$

$$t = \frac{|0.086\ 86 - 0.086\ 41|}{0.001\ 08}\sqrt{\frac{18 \cdot 29}{18 + 29}} = 1.39$$

For 47 degrees of freedom $t_{\text{table}} \approx 2.01$.

$t_{\text{calculated}} = 1.39 < t_{\text{table}}$, so the difference between means is <u>not</u> significant.

4-11. $F_{\text{calculated}} = \dfrac{s_1^2}{s_2^2} = \dfrac{10^2}{8^2} = 1.5_6$

For 4 degrees of freedom in the numerator and 3 in the denominator, $F_{\text{table}} = 9.12$.
$F_{\text{calculated}} < F_{\text{table}}$, so standard deviations are <u>not</u> significantly different.

$$s_{\text{pooled}} = \sqrt{\dfrac{8^2(4-1) + 10^2(5-1)}{4+5-2}} = 9.20$$

$$t_{\text{calculated}} = \dfrac{|238 - 255|}{9.20}\sqrt{\dfrac{4 \cdot 5}{4+5}} = 2.75$$

$t_{\text{table}} = 2.365$ for 95% confidence and $4+5-2 = 7$ degrees of freedom

$t_{\text{calculated}} > t_{\text{table}}$, so the difference <u>is</u> significant at the 95% confidence level.

4-12. For Method 1, we compute $\bar{x}_1 = 0.082\,605_2$, $s_1 = 0.000\,013_4$

For Method 2, $\bar{x}_2 = 0.082\,573_2$, $s_2 = 0.000\,026_0$

$$F_{\text{calculated}} = \dfrac{s_1^2}{s_2^2} = \dfrac{0.000\,026_0^2}{0.000\,013_4^2} = 3.7_6$$

For 5 degrees of freedom in the numerator and 4 in the denominator, $F_{\text{table}} = 6.26$.
$F_{\text{calculated}} < F_{\text{table}}$, so standard deviations are <u>not</u> significantly different.

$$s_{\text{pooled}} = \sqrt{\dfrac{0.000\,013_4^2(5-1) + 0.000\,026_0^2(6-1)}{5+6-2}} = 0.000\,021_3$$

$$t_{\text{calculated}} = \dfrac{|0.082\,605_2 - 0.082\,573_2|}{0.000\,021_3}\sqrt{\dfrac{5 \cdot 6}{5+6}} = 2.48$$

$t_{\text{table}} = 2.262$ for 95% confidence and $5+6-2 = 9$ degrees of freedom

$t_{\text{calculated}} > t_{\text{table}}$, so the difference <u>is</u> significant at the 95% confidence level.

4-13. (a) $F_{\text{calculated}} = \dfrac{s_1^2}{s_2^2} = \dfrac{0.09^2}{0.02^2} = 20._2$

For 5 degrees of freedom in the numerator and denominator, $F_{\text{table}} = 5.05$.
$F_{\text{calculated}} > F_{\text{table}}$, so standard deviations <u>are</u> significantly different.

$$t_{\text{calculated}} = \dfrac{|\bar{x}_1 - \bar{x}_2|}{\sqrt{s_1^2/n_1 + s_2^2/n_2}} = \dfrac{|10.01 - 9.98|}{\sqrt{0.09^2/6 + 0.02^2/6}} = 0.80$$

$$\text{Degrees of freedom} = \frac{(s_1^2/n_1 + s_2^2/n_2)^2}{\dfrac{(s_1^2/n_1)^2}{n_1-1} + \dfrac{(s_2^2/n_2)^2}{n_2-1}}$$

$$= \frac{(0.09^2/6 + 0.02^2/6)^2}{\dfrac{(0.09^2/6)^2}{6-1} + \dfrac{(0.02^2/6)^2}{6-1}} = 5.49 \approx 5$$

For 5 degrees of freedom, the critical value of t for 95% confidence is 2.571. $t_{\text{calculated}} = 0.80 < t_{\text{table}}$, so the difference in means is not significant.

(b) $F_{\text{calculated}} = \dfrac{s_1^2}{s_2^2} = \dfrac{0.09^2}{0.03^2} = 9._{00}$

For 5 degrees of freedom in the numerator and denominator, $F_{\text{table}} = 5.05$. $F_{\text{calculated}} > F_{\text{table}}$, so standard deviations are significantly different.

$$t_{\text{calculated}} = \frac{|\bar{x}_1 - \bar{x}_2|}{\sqrt{s_1^2/n_1 + s_2^2/n_2}} = \frac{|10.01 - 9.80|}{\sqrt{0.09^2/6 + 0.03^2/6}} = 5.4_2$$

$$\text{Degrees of freedom} = \frac{(s_1^2/n_1 + s_2^2/n_2)^2}{\dfrac{(s_1^2/n_1)^2}{n_1-1} + \dfrac{(s_2^2/n_2)^2}{n_2-1}}$$

$$= \frac{(0.09^2/6 + 0.03^2/6)^2}{\dfrac{(0.09^2/6)^2}{6-1} + \dfrac{(0.03^2/6)^2}{6-1}} = 6.10 \approx 6$$

For 6 degrees of freedom, the critical value of t for 95% confidence is 2.447. $t_{\text{calculated}} = 5.4_2 > t_{\text{table}}$, so the difference in means is significant.

(c) $F_{\text{calculated}} = \dfrac{s_1^2}{s_2^2} = \dfrac{0.02^2}{0.009^2} = 4._{94}$

For 5 degrees of freedom in the numerator and denominator, $F_{\text{table}} = 5.05$. $F_{\text{calculated}} < F_{\text{table}}$, so standard deviations are not significantly different.

$$s_{\text{pooled}} = \sqrt{\frac{0.02^2(6-1) + 0.009^2(6-1)}{6+6-2}} = 0.01_{55}$$

$$t_{\text{calculated}} = \frac{|9.98 - 10.004|}{0.01_{55}} \sqrt{\frac{6 \cdot 6}{6+6}} = 2.6_8$$

$t_{\text{table}} = 2.228$ for 95% confidence and $6 + 6 - 2 = 10$ degrees of freedom. $t_{\text{calculated}} > t_{\text{table}}$, so the difference is significant at the 95% confidence level.

(d) The buret and pipet have similar accuracy, but the flask delivers a low value. Tolerances for Class A glassware are found in the tables in Chapter 2:

50-mL buret: ±0.05 mL   10-mL pipet: ±0.02 mL   10-mL flask: ±0.02 mL

The observed accuracy for the buret and pipet are within the manufacturer's tolerance. The flask is far from the manufacturer's tolerance. Either the flask is out of tolerance or neither student knows how to read the level of the meniscus in the flask correctly.

4-14. $\bar{x} = 0.2168$; $s = 0.0110$

$G_{calculated} = |0.195 - 0.2168| / 0.0110 = 1.98 > G_{table} (= 1.822)$

⇒ Discard 0.195

4-15. $\bar{x} = 0.1819$; $s = 0.0057$;   Outlier = 0.169

$G_{calculated} = |0.169 - 0.1819| / 0.0057 = 2.26 > G_{table} (= 2.176)$

⇒ Discard 0.169

4-16. $\dfrac{y_2 - y_1}{x_2 - x_1} = \dfrac{y - y_1}{x - x_1}$ $\dfrac{-1 - 3}{8 - 6} = \dfrac{y - 3}{x - 6}$

$-4(x - 6) = 2(y - 3) \Rightarrow y = -2x + 15$

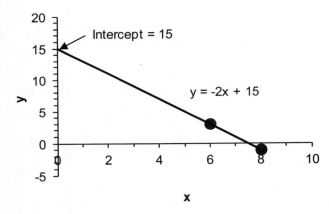

4-17. The first uncertain figure is the last significant figure.

Slope $= -1.29872 \times 10^4$ ($\pm 0.0013190 \times 10^4$) $= -1.299 (\pm 0.001) \times 10^4$

Intercept $= 256.695 (\pm 323.57) = 3 (\pm 3) \times 10^2$

4-18. (a) $x = \dfrac{y-b}{m} = \dfrac{2.58 - 1.34615}{0.61538} = 2.005$

$\bar{x} = (\Sigma x_i)/n = (1 + 3 + 4 + 6)/4 = 3.50$

$\bar{y} = (\Sigma y_i)/n = (2 + 3 + 4 + 5)/4 = 3.50$

$\Sigma(x_i - \bar{x})^2 = (1 - 3.5)^2 + (3 - 3.5)^2 + (4 - 3.5)^2 + (6 - 3.5)^2 = 13$

$s_x = \dfrac{s_y}{|m|}\sqrt{\dfrac{1}{k} + \dfrac{1}{n} + \dfrac{(y-\bar{y})^2}{m^2 \Sigma(x_i - \bar{x})^2}}$

$= \dfrac{0.19612}{0.61538}\sqrt{\dfrac{1}{1} + \dfrac{1}{4} + \dfrac{(2.58 - 3.50)^2}{0.61538^2\,(13)}} = 0.38$

Answer: $2.0_0 \pm 0.3_8$

(b) Replace the first term in the square root $\left(\dfrac{1}{1}\right)$ by $\dfrac{1}{4} \Rightarrow s_x = 0.26$

Answer: $2.0_0 \pm 0.2_6$

4-19. (a) and (b)

	A	B	C	D
1	Least-Squares Spreadsheet			
2			Corrected	Observed
3		x	y	y
4	Highlight cells B12:C14	0	0.000	0.466
5	Type "= LINEST(C4:C8,	9.36	0.210	0.676
6	B4:B8,TRUE,TRUE)	18.72	0.417	0.883
7	For PC, press	28.08	0.620	1.086
8	CTRL+SHIFT+ENTER	37.44	0.814	1.280
9	For Mac, press			
10	APPLE+ENTER			
11			LINEST output:	
12		m	0.02177	0.00460 b
13		s_m	0.00019	0.00439 s_b
14		R^2	0.99977	0.00567 s_y
15				
16	n =		5	B16 = COUNT(B4:B8)
17	Mean y =		0.412	B17 = AVERAGE(C4:C8)
18	$\Sigma(x_i$ - mean $x)^2$ =		876.096	B28 = DEVSQ(B4:B8)
19				
20	Measured y =		0.507	Input = observed absorbance - blank
21	k = number of replicate measurements of y =		1	Input
22	Derived x		23.0739	B22 = (B20-C12)/B12
23	s_x =		0.2878	B23 = (C14/B12)*SQRT((1/B21)+(1/B16)+((B20-B17)^2)/(B12^2*B18))

(c) The results for unknown protein are in cells B22 and B23 in the spreadsheet:
unknown protein = $23.0_7 \pm 0.2_9$ μg

4-20. (a),(b) The following spreadsheet covers half of the required range.

	A	B	C	D	E
1	Gaussian Curve				
2					
3	Constants:	x	y (s = 1)	y (s = 2)	
4	Average =	4	6.08E-09	2.22E-03	
5	10	4.5	1.08E-07	4.55E-03	
6	StdDev =	5	1.49E-06	8.76E-03	
7	1	5.5	1.60E-05	1.59E-02	
8	2*pi =	6	1.34E-04	2.70E-02	
9	6.283185	6.5	8.73E-04	4.31E-02	
10		7	4.43E-03	6.48E-02	
11		7.5	1.75E-02	9.13E-02	
12		8	5.40E-02	1.21E-01	
13	C4 =				
14	(1/($A$7*SQRT($A$9)))*EXP(-((B4-$A$5)^2)/(2*$A$7^2))				

(c)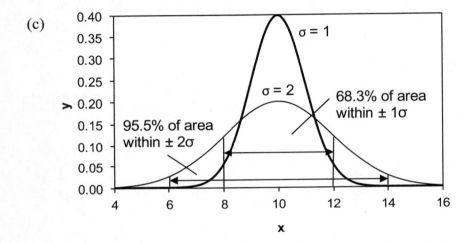

4-21. The formula mass of $CuCO_3$ is 123.555 and Cu in this formula is 51.43 wt%. For the 1995 class data, the 95% confidence interval is

$$\mu\,(95\%) = \bar{x} \pm \frac{ts}{\sqrt{n}} = 55.6 \pm \frac{(2.02)(2.7)}{\sqrt{43}} = 55.6 \pm 0.8 = 54.8 \text{ to } 56.4 \text{ wt}\%$$

The 99% confidence interval is

$$\mu\,(99\%) = \bar{x} \pm \frac{ts}{\sqrt{n}} = 55.6 \pm \frac{(2.70)(2.7)}{\sqrt{43}} = 55.6 \pm 1.1 = 54.5 \text{ to } 56.7 \text{ wt}\%$$

Even the 99% confidence interval does not include the Cu content in $CuCO_3$ (51.43 wt%). For the 1996 class data, the 99% confidence interval is 54.3 to 57.5 wt%. From the instructor's measurements, the 99% confidence interval is 55.2 to 56.4 wt%. The product cannot be $CuCO_3$. It cannot be a hydrate either, because

$CuCO_3 \cdot xH_2O$, would have an even lower Cu content than 51.43%. The observed composition is closer to that of the minerals azurite, $Cu_3(OH)_2(CO_3)_2$ (55.31 wt% Cu), or malachite, $Cu_2(OH)_2(CO_3)$ (57.48 wt% Cu), than it is to $CuCO_3$.

4-22. The 95% confidence intervals for the EH isotope ratios $^{87}Sr/^{86}Sr$ from Dome C (0.706 2 to 0.707 4) and Law Dome (0.709 3 to 0.710 1) do not overlap, so it is likely that the particles at these locations came from different sources. The 95% confidence intervals for the LGM isotope ratios $^{87}Sr/^{86}Sr$ from Dome C (0.707 7 to 0.708 7) and Law Dome (0.708 2 to 0.710 4) do overlap, so the dust at these two sites could have come from the same source.

The unit pg/g means 10^{-12} grams of strontium per gram of ice. A descriptive term for pg/g is parts per trillion.

CHAPTER 5
QUALITY ASSURANCE AND CALIBRATION METHODS

5-1. Raw data are individual values of a measured quantity, such as peak areas from a chromatogram or volumes from a buret. Treated data are concentrations or amounts found by applying a calibration method to the raw data. Results, such as the mean and standard deviation, are what we ultimately report after applying statistics to treated data.

5-2. A calibration check is an analysis of a solution *formulated by the analyst* to contain a known concentration of analyte. It is the analyst's own check that procedures and instruments are functioning correctly. A performance test sample is an analysis of a solution *formulated by someone other than the analyst* to contain a known concentration of analyte. It is a test to see whether the analyst gets correct results when he or she does not know what the correct result should be.

5-3. A blank is a sample intended to contain no analyte. Its purpose is to find the response of a method when no analyte is deliberately present. A positive analytical response to the blank arises from analyte impurities in reagents and equipment and from interference by other species. A method blank is taken through all steps in a chemical analysis. A reagent blank is the same as a method blank, but it has not been subjected to all sample preparation procedures. A field blank is similar to a method blank, but it has been taken into the field and exposed to the same environment as samples collected in the field and transported to the lab.

5-4. Linear range is the analyte concentration interval over which the analytical signal is proportional to analyte concentration. Dynamic range is the concentration range over which there is a measurable (nonzero) response to analyte, even if it is not linear. Range is the analyte concentration interval over which an analytical method has specified linearity, accuracy, and precision.

5-5. A false positive is a conclusion that the concentration of analyte exceeds a certain limit when, in fact, the concentration is below the limit. A false negative is a conclusion that the concentration of analyte is below a certain limit when, in fact, the concentration is above the limit.

5-6. ~1% of the area under the curve for blanks lies to the right of the detection limit. Therefore, ~1% of samples containing no analyte will give a signal above the detection limit. 50% of the area under the curve for samples containing analyte at the detection limit lies below (to the left) of the detection limit. Therefore, 50% of samples containing analyte at the detection limit will be reported as not containing analyte at a level above the detection limit.

5-7. A control chart tracks the performance of a process to see whether it remains within expected bounds. Six indications that a process might be out of control are (1) a reading outside the action lines, (2) 2 out of 3 consecutive readings between the warning and action lines, (3) 7 consecutive measurements all above or all below the center line, (4) 6 consecutive measurements all steadily increasing or all steadily decreasing, wherever they are located, (5) 14 consecutive points alternating up and down, regardless of where they are located, and (6) an obvious nonrandom pattern.

5-8. Statement **(c)** is correct. The purpose of the analysis is to see whether concentrations of haloacetates are in compliance with (i.e., do not exceed) levels set by a certain rule. The purpose is not just to achieve a specified precision and accuracy (statement **(a)**) or to just see whether detectable levels of haloacetates are present (statement **(b)**).

5-9. 50% of red wells should be green and 8% of green wells should be red. Therefore, half the wells labeled as unsafe are really safe, so a great deal of usable water is not used. If the false negative rate were 50%, then 50% of the wells labeled as safe would really not be safe—a much worse kind of error.

5-10. $F_{\text{calculated}} = \dfrac{s_1^2}{s_1^2} = \dfrac{0.78^2}{0.56^2} = 1.9_4$

For 2 degrees of freedom in the numerator and 7 in the denominator, $F_{\text{table}} = 4.74$. $F_{\text{calculated}} < F_{\text{table}}$, so the standard deviations are <u>not</u> significantly different.

We use the t test with Equations 4-6 and 4-5 to compare results:

$$s_{\text{pooled}} = \sqrt{\dfrac{s_1^2(n_1-1)+s_1^2(n_2-1)}{n_1+n_2-2}} = \sqrt{\dfrac{0.56^2(8-1)+0.78^2(3-1)}{8+3-2}} = 0.61_6$$

$$t_{\text{calculated}} = \dfrac{|\bar{x}_1 - \bar{x}_2|}{s_{\text{pooled}}}\sqrt{\dfrac{n_1 n_2}{n_1+n_2}} = \dfrac{|1.65-2.68|}{0.61_6}\sqrt{\dfrac{8 \cdot 3}{8+3}} = 2.4_7$$

$t_{\text{table}} = 2.262$ for 95% confidence and $8+3-2 = 9$ degrees of freedom
$t_{\text{calculated}} > t_{\text{table}}$, so the difference <u>is</u> significant at the 95% confidence level.

5-11. (a) Standard deviation of 9 samples: $s = 0.000\,6_{44}$

Mean blank: $y_{blank} = 0.001\,1_{89}$

Signal limit of detection $= y_{blank} + 3s$
$= 0.001\,1_{89} + (3)(0.000\,6_{44}) = 0.003_{12}$

(b) Concentration detection limit $= \dfrac{3s}{m} = \dfrac{(3)(0.000\,6_{44})}{2.24 \times 10^4 \text{ M}^{-1}} = 8.6 \times 10^{-8}$ M

(c) Lower limit of quantitation $= \dfrac{10s}{m} = \dfrac{(10)(0.000\,6_{44})}{2.24 \times 10^4 \text{ M}^{-1}} = 2.9 \times 10^{-7}$ M

5-12. Mean $= 0.84_1$ ppb; Standard deviation $= 0.18_9$ ppb

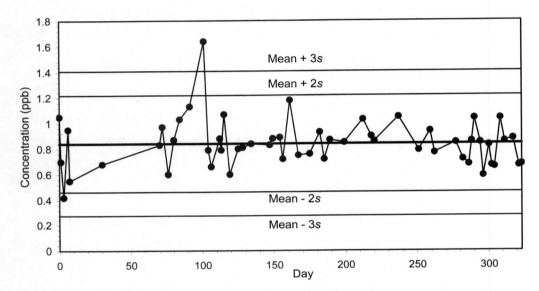

Stability criteria are

- No observations outside the action lines—one observation (day 101) lies above the upper action line.
- There are not 2 out of 3 consecutive measurements between the warning and action lines—OK
- There are not 7 consecutive measurements all above or all below the center line—OK
- There are not 6 consecutive measurements all steadily increasing or all steadily decreasing, wherever they are located—OK
- There are not 14 consecutive points alternating up and down, regardless of where they are located—OK
- There is no obvious nonrandom pattern—OK

The only criterion that was violated was one observation outside the action line.

5-13. Criteria:
- Observations outside action lines—no
- 2 out of 3 consecutive measurements between warning and action lines—no
- 7 consecutive measurements all above or all below the center line—YES: Observations 2–10 (starting from the left side) are all below the center line and observations 27–33 are all above the center line.
- 6 consecutive measurements steadily increasing or steadily decreasing—no
- 14 consecutive points alternating up and down—no
- Obvious nonrandom pattern—no

5-14. (a) For the fortification level of 22.2 ng/mL, the mean of the 5 values is 23.6_6 ng/mL and the standard deviation is 5.6_3 ng/mL.

$$\text{Precision} = 100 \times \frac{5.63}{23.66} = 23.8\%$$

$$\text{Accuracy} = 100 \times \frac{23.66 - 22.2}{22.2} = 6.6\%$$

For the fortification level of 88.2 ng/mL, the mean of the 5 values is 82.4_8 ng/mL and the standard deviation is 11.4_9 ng/mL.

$$\text{Precision} = 100 \times \frac{11.49}{82.48} = 13.9\%$$

$$\text{Accuracy} = 100 \times \frac{82.48 - 88.2}{88.2} = -6.5\%$$

For the fortification level of 314 ng/mL, the mean of the 5 values is 302.8 ng/mL and the standard deviation is 23.5_1 ng/mL.

$$\text{Precision} = 100 \times \frac{23.51}{302.8} = 7.8\%$$

$$\text{Accuracy} = 100 \times \frac{302.8 - 314}{314} = -3.6\%$$

(b) Standard deviation of 10 samples: $s = 28._2$; mean blank: $y_{\text{blank}} = 45._0$

Signal detection limit $= y_{\text{blank}} + 3s = 45._0 + (3)(28._2) = 129._6$

Concentration detection limit $= \dfrac{3s}{m} = \dfrac{(3)(28._2)}{1.75 \times 10^9 \text{ M}^{-1}} = 4.8 \times 10^{-8}$ M

Lower limit of quantitation $= \dfrac{10s}{m} = \dfrac{(10)(28._2)}{1.75 \times 10^9 \text{ M}^{-1}} = 1.6 \times 10^{-7}$ M

5-15. Mean = 0.383 µg/L;   standard deviation = 0.0214 µg/L

% recovery = $\dfrac{0.383 \text{ µg/L}}{0.40 \text{ µg/L}} \times 100 = 96\%$

The measurements are already expressed in concentration units. The concentration detection limit is 3 times the standard deviation = $3(0.0214 \text{ µg/L})$ = 0.064 µg/L.

5-16. If the initial concentration of vitamin C in the juice is $[X]_i$, the concentration after dilution of 50.0 mL of juice with 1.00 mL of standard is

$$\text{Final concentration of analyte} = [X]_f = \frac{\text{initial volume}}{\text{final volume}}[X]_i = \frac{50.0 \text{ mL}}{51.0 \text{ mL}}[X]_i$$

The final concentration of the added standard after addition to the juice is

$$[S]_f = \frac{1.00}{51.0}[S]_i = \frac{1.00}{51.0}[29.4 \text{ mM}] = 0.576_5 \text{ mM}$$

The standard addition equation therefore becomes

$$\frac{[X]_i}{[X]_f + [S]_f} = \frac{[X]_i}{\dfrac{50.0}{51.0}[X]_i + 0.576_5 \text{ mM}} = \frac{2.02 \text{ µA}}{3.79 \text{ µA}} \Rightarrow [X]_i = 0.644 \text{ mM}$$

5-17. $\dfrac{[X]_i}{[S]_f + [X]_f} = \dfrac{I_X}{I_{S+X}}$

$$\frac{[X]_i}{\left(\dfrac{1.00 \text{ mL}}{101.0 \text{ mL}}\right)(0.0500 \text{ M}) + \left(\dfrac{100.0 \text{ mL}}{101.0 \text{ mL}}\right)[X]_i} = \frac{10.0 \text{ mV}}{14.0 \text{ mV}} \Rightarrow [X]_i = 1.21 \text{ mM}$$

5-18. (a) Figure 5-7 is the appropriate graph because all solutions were made to the same total volume. Figure 5-6 would be used for successive standard additions that increase the volume of the solution for every addition. In Figure 5-7, the ordinate is the observed signal I_{S+X} and the abscissa is the concentration of diluted standard, $[S]_f$, in each solution. In Figure 5-6, the ordinate is $I_{S+X}(V/V_0)$ and the abscissa is $[S]_i(V_s/V_0)$. In Figure 5-7, the negative x-intercept is the final concentration of analyte in the constant-volume samples that were analyzed. In Figure 5-6, the negative x-intercept is the concentration of analyte in the original unknown.

(b) All solutions were made up to the same final volume, so we prepare a graph of signal versus concentration of added standard. The line in the graph was drawn by the method of least squares, and the x-intercept is –8.72 ppb.

Quality Assurance and Calibration Methods

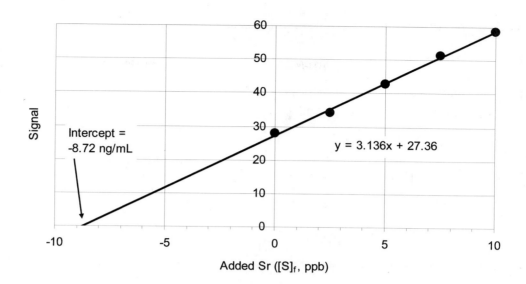

(c) The unknown solution has a volume of 10.0 mL with a Sr concentration of 8.72 ppb = 8.72 ng/mL. In 10.0 mL, there are (10 mL)(8.72 ng/mL) = 87.2 ng. The solution was made from 0.750 mg of tooth enamel. The concentration of Sr in tooth enamel, in ppm, is

$$\text{Concentration (ppm)} = \frac{\text{mass of Sr}}{\text{mass of enamel}} \times 10^6$$

$$= \frac{87.2 \times 10^{-9} \text{ g}}{0.750 \times 10^{-3} \text{ g}} \times 10^6 = 116 \text{ ppm}$$

5-19. (a) In the spreadsheet below, the x-intercept is given by $-b/m$ in cell B20 = -8.72 ppb. The uncertainty (standard deviation) in intercept is given in cell B27 (0.43 ppb) using the formula in row 28.

Sr concentration in 10 mL solution = 8.72 ± 0.43 ppb

(b) $\text{Concentration (ppm)} = \dfrac{\text{mass of Sr}}{\text{mass of enamel}} \times 10^6$

$$= \frac{87.2 (\pm 4.3) \times 10^{-9} \text{ g}}{0.750 \times 10^{-3} \text{ g}} \times 10^6 = \frac{87.2 (\pm 4.9\%) \times 10^{-9} \text{ g}}{0.750 \times 10^{-3} \text{ g}} \times 10^6$$

$$= 116 (\pm 4.9\%) \text{ ppm} = 116 \pm 6 \text{ ppm}$$

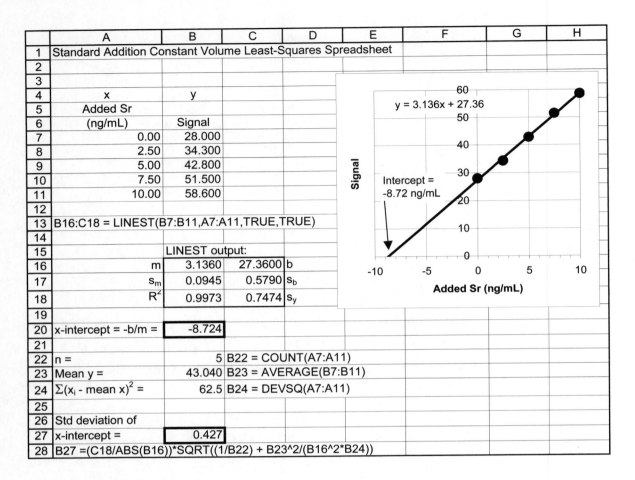

5-20. All solutions have the same volume, so we prepare a graph of I vs $[S]_f$. The intercept is at -313 ppb, so the original concentration of Cu^{2+} is 313 ppb.

Added standard (ppb)	Current measured in figure (μA)
0	0.59$_9$
100	0.77$_4$
200	0.94$_3$
300	1.12$_8$
400	1.31$_4$
500	1.54$_4$

Quality Assurance and Calibration Methods 37

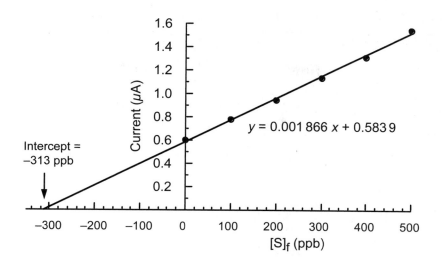

5-21. (a) In the spreadsheet, data appear in columns B and D. The abscissa and ordinate functions are in columns C and E.

	A	B	C	D	E
1	Standard Addition Experiment				
2	Add 2.50 ppm standard Pb to river sediment extract				
3					
4		V_S =			y-axis function
5	V_0 (mL) =	mL Pb std	x-axis function	I_{S+X} =	$I_{S+X} \cdot V/V_0$
6	4.60	added	$[S]_i \cdot V_S/V_0$ (ppm)	signal	($V = V_0 + V_S$)
7	$[S]_i$ (ppm) =	0.000	0	1.10	1.100
8	2.50	0.025	0.0136	1.66	1.669
9		0.050	0.0272	2.20	2.224
10		0.075	0.0408	2.81	2.856
11	V = total volume = $V_0 + V_S$				
12					
13			C17:D19 = LINEST(E7:E10,C7:C10,TRUE,TRUE)		
14					
15			LINEST output:		
16			m	42.8524	1.0888	b
17			s_m	0.8720	0.0222	s_b
18			R^2	0.9992	0.0265	s_y
19					
20			x-intercept	-0.02541	
21					
22			n =	4	C22 = COUNT(C7:C10)
23			Mean y =	1.962	C23 = AVERAGE(E7:E10)
24		$\Sigma(x_i - \text{mean } x)^2$ =		0.000923027	C24 = DEVSQ(C7:C10)
25					
26			Std deviation of		
27			x-intercept	0.00098	
28	C27 = (D18/ABS(C16))*SQRT((1/C22) + C23^2/(C16^2*C24))				

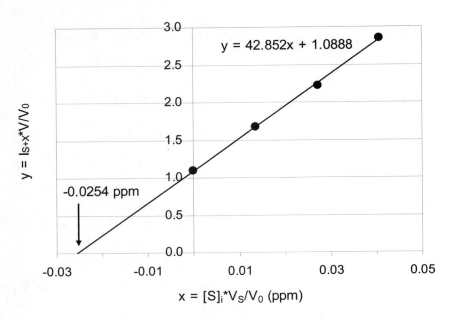

The negative x-intercept (0.025 4 ppm) is the concentration of Pb(II) in the initial sample volume $V_0 = 4.60$ mL. The concentration of Pb(II) in the 1.00-mL extract is found from the dilution formula:

$(x \text{ ppm})(1.00 \text{ mL}) = (0.025\ 4 \text{ ppm})(4.60 \text{ mL}) \Rightarrow x = 0.116\ 9$ ppm

(b) The uncertainty in the intercept in cell C27 is 0.000 98 ppm. The relative uncertainty in the intercept is $100 \times (0.000\ 98 \text{ ppm}/0.025\ 4 \text{ ppm}) = 3.86\%$. If other sources of uncertainty are negligible, the relative uncertainty in concentration in the 1.00-mL extract is also 3.86%, thus giving an absolute uncertainty of $(0.038\ 6)(0.116\ 9 \text{ ppm}) = 0.004\ 52$ ppm. A reasonable expression of the answer is $0.116_9 \pm 0.004_5$ ppm or 0.117 ± 0.005 ppm.

5-22. First evaluate the response factor from the known mixture:

$$\frac{A_X}{[X]} = F\left(\frac{A_S}{[S]}\right)$$

$$\frac{306}{12.8\ \mu M} = F\left(\frac{511}{44.4\ \mu M}\right) \Rightarrow F = 2.07_7$$

For the unknown mixture, we can write

$$\frac{A_X}{[X]} = F\left(\frac{A_S}{[S]}\right)$$

$$\frac{251}{[X]} = 2.07_7 \left(\frac{563}{55.5\ \mu M}\right) \Rightarrow [X] = 11.9\ \mu M$$

5-23. [S] in unknown mixture $= (8.24\ \mu g/mL)\left(\dfrac{5.00\ mL}{50.0\ mL}\right) = 0.824\ \mu g/mL$

Standard mixture has equal concentrations of X and S:

$$\dfrac{A_X}{[X]} = F\left(\dfrac{A_S}{[S]}\right) \Rightarrow \dfrac{0.930}{[\cancel{X}]} = F\left(\dfrac{1.000}{[\cancel{S}]}\right) \Rightarrow F = 0.930$$

Unknown mixture:

$$\dfrac{A_X}{[X]} = F\left(\dfrac{A_S}{[S]}\right) \Rightarrow \dfrac{1.69}{[X]} = 0.930\left(\dfrac{1.000}{[0.824\ \mu g/mL]}\right)$$

$$\Rightarrow [X] = 1.49_7\ \mu g/mL$$

But X was diluted by a factor of 10.00 mL/50.0 mL, so the original concentration in the unknown was

$(1.49_7\ \mu g/mL)\left(\dfrac{50.0\ mL}{10.00\ mL}\right) = 7.49\ \mu g/mL$

5-24. S = pentanol and X = hexanol. We will substitute mmol for concentrations because the volume is unknown and concentration is proportional to the number of mmol in the sample.

Standard mixture:

$$\dfrac{A_X}{[X]} = F\left(\dfrac{A_S}{[S]}\right) \Rightarrow \dfrac{1\,570}{[1.53]} = F\left(\dfrac{922}{[1.06]}\right) \Rightarrow F = 1.18_0$$

Unknown mixture:

$$\dfrac{A_X}{[X]} = F\left(\dfrac{A_S}{[S]}\right) \Rightarrow \dfrac{816}{[X]} = 1.18_0\left(\dfrac{843}{[0.57]}\right) \Rightarrow [X] = 0.47\ mmol$$

5-25. If an athlete tests positive for drugs, the test should be repeated with a second sample that was drawn at the same time as the first sample and preserved in an appropriate manner. If there is a 1% chance of a false positive in each test, the chances of observing a false positive twice in a row are 1% of 1% or 0.01%. Instead of falsely accusing 1% of innocent athletes, we would be falsely accusing 0.01% of innocent athletes.

CHAPTER 6
GOOD TITRATIONS

6-1. The equivalence point occurs when the exact stoichiometric quantities of reagents have been mixed. The end point, which comes near the equivalence point, is marked by a sudden change in a physical property brought about by the disappearance of a reactant or appearance of a product.

6-2. 108.0 mL of 0.1650 M oxalic acid = (0.1080 L)(0.1650 M) = 0.01782 mol oxalic acid, which requires

$$\left(\frac{2 \text{ mol MnO}_4^-}{5 \text{ mol H}_2\text{C}_2\text{O}_4}\right)(0.01782 \text{ mol H}_2\text{C}_2\text{O}_4) = 7.128 \times 10^{-3} \text{ mol MnO}_4^-$$

7.128 mmol / (0.1650 mmol/mL) = 43.2 mL of $KMnO_4$

An easy way to see this is to note that the reagents are both 0.1650 M.

$$\text{volume of MnO}_4^- = \tfrac{2}{5}(\text{volume of oxalic acid})$$

For the second part of the question,

$$\text{volume of oxalic acid} = \tfrac{5}{2}(\text{volume of MnO}_4^-) = 270.0 \text{ mL}$$

6-3. Subtract the blank volume from the end-point volume to find the equivalence volume: 15.44 – 0.04 = 15.40 mL. This much titrant solution contains mol MnO_4^- = (0.01117 M)(0.01540 L) = $1.720_2 \times 10^{-4}$ mol. The moles of oxalic acid in the unknown must be

$$\text{mol H}_2\text{C}_2\text{O}_4 = (1.720_2 \times 10^{-4} \text{ mol MnO}_4^-)\left(\frac{5 \text{ mol H}_2\text{C}_2\text{O}_4}{2 \text{ mol MnO}_4^-}\right) = 4.300_4 \times 10^{-4} \text{ mol}$$

The concentration of oxalic acid in the unknown is

$$[\text{H}_2\text{C}_2\text{O}_4] = \frac{4.300_4 \times 10^{-4} \text{ mol}}{0.01000 \text{ L}} = 4.300 \times 10^{-2} \text{ M}$$

6-4. 1.69 mg of NH_3 = 9.92×10^{-5} mol of NH_3. This quantity will react with

$$\left(\frac{3 \text{ mol OBr}^-}{2 \text{ mol NH}_3}\right)(9.92 \times 10^{-5} \text{ mol of NH}_3) = 1.49 \times 10^{-4} \text{ mol of OBr}^-$$

The molarity of OBr^- is 0.149 mmol/1.00 mL = 0.149 M.

6-5. 40.0 mL of 0.0400 M $Hg_2(NO_3)_2$ = (0.0400 L)(0.0400 M) = 1.60×10^{-3} mol of Hg_2^{2+}, which will require 3.20×10^{-3} mol of KI.

This quantity is contained in volume = $\dfrac{3.20 \times 10^{-3} \text{ mol}}{0.100 \text{ mol/L}}$ = 32.0 mL

6-6. (a) $\text{mol Hg}^{2+} = \frac{1}{2} \text{mol Cl}^- = \frac{1}{2}\left(\frac{0.1476 \text{ g}}{58.44 \text{ g/mol}}\right) = 1.263 \times 10^{-3}$ mol

$$[\text{Hg}^{2+}] = \frac{1.263 \times 10^{-3} \text{ mol}}{0.028\,06 \text{ L}} = 0.045\,00 \text{ M}$$

(b) $(0.022\,83 \text{ L Hg(NO}_3)_2)(0.045\,00 \text{ M}) = 1.027 \times 10^{-3}$ mol Hg^{2+}
   $= 2.055 \times 10^{-3}$ mol Cl^-
   $(2.055 \times 10^{-3} \text{ mol Cl}^-)(35.453 \text{ g Cl}^-/\text{mol Cl}^-) = 72.85$ mg Cl^-
   $(72.85 \text{ mg Cl}^-)/(2.000 \text{ mL}) = 36.42$ mg Cl^-/mL

6-7. 50.00 mL of 0.365 0 M $\text{AgNO}_3 = 18.25 \times 10^{-3}$ mol of Ag^+
37.60 mL of 0.287 0 M KSCN $= 10.79 \times 10^{-3}$ mol of SCN^-
Difference $= 18.25 - 10.79 = 7.46 \times 10^{-3}$ mol of $\text{I}^- = 947$ mg of I^-

6-8. 1.00 mL of 0.027 3 M $\text{Ce}^{4+} = 0.0273$ mmol of Ce^{4+}

This quantity will react with half as many mmol of oxalic acid $= 0.013\,65$ mmol of $\text{H}_2\text{C}_2\text{O}_4 \cdot 2\text{H}_2\text{O} = 1.72$ mg.

6-9. (a) Theoretical molarity $= (3.214 \text{ g}/158.03 \text{ g/mol})/(1.00 \text{ L}) = 0.020\,34$ M

(b) 25.00 mL of 0.020 34 M $\text{KMnO}_4 = 0.508\,5 \times 10^{-3}$ mol. But 2 mol MnO_4^- reacts with 5 mol H_3AsO_3, which comes from 2.5 mol As_2O_3 (because 1 As_2O_3 provides 2 H_3AsO_3).

Moles of As_2O_3 needed to react with 0.508 5 mmol of MnO_4^-:

$$\left(\frac{1 \text{ mol As}_2\text{O}_3}{2 \text{ mol H}_3\text{AsO}_3}\right)\left(\frac{5 \text{ mol H}_3\text{AsO}_3}{2 \text{ mol KMnO}_4}\right)(0.508\,5 \times 10^{-3} \text{ mol KMnO}_4) =$$

$0.635\,6 \times 10^{-3}$ mol $\text{As}_2\text{O}_3 = 0.125\,7$ g of As_2O_3

(c) $\dfrac{0.508\,5 \text{ mmol KMnO}_4}{0.125\,7 \text{ g As}_2\text{O}_3} = \dfrac{x \text{ mmol KMnO}_4}{0.146\,8 \text{ g As}_2\text{O}_3}$

$x = 0.593\,9$ mmol KMnO_4 in $(29.98 - 0.03) = 29.95$ mL
$[\text{KMnO}_4] = 0.593\,9$ mmol/29.95 mL $= 0.019\,83$ M

6-10. (a) $\dfrac{0.824 \text{ g acid}}{204.22 \text{ g/mol}} = 4.03_5$ mmol This quantity of NaOH is contained in 0.038 314 kg of NaOH solution.

concentration of NaOH $= \dfrac{4.03_5 \times 10^{-3} \text{ mol NaOH}}{0.038\,314 \text{ kg solution}}$

$= 0.105\,3$ mol/kg solution

(b) mol NaOH = (0.057 911 kg)(0.105 3 mol/kg) = 6.09$_9$ mmol

Two moles NaOH react with 1 mol H$_2$SO$_4$, so mol H$_2$SO$_4$ = 6.09$_9$ mmol/2
= 3.04$_9$ mmol

$$[H_2SO_4] = \frac{3.04_9 \text{ mmol}}{0.010\ 63 \text{ kg}} = 0.286_9 \text{ mol H}_2\text{SO}_4/\text{kg solution}$$

6-11. (a) mol Ag$^+$ = $\dfrac{4.872 \pm 0.003 \text{ g}}{169.873\ 1 \text{ g/mol}} = \dfrac{4.872 \pm 0.061_6\% \text{ g}}{169.873\ 1 \pm 0\% \text{ g/mol}}$

= 0.028 680 ($\pm 0.061_6\%$) mol

(propagation of uncertainty for division: $\%e = \sqrt{\%e_1^2 + \%e_2^2}$)

mass of solution = 498.633 $\pm$ 0.003 g H$_2$O + 4.872 $\pm$ 0.003 g AgNO$_3$

= 503.505 $\pm$ 0.004$_{24}$ g = 503.505 $\pm$ 0.000 0843%

(propagation of uncertainty for addition: $0.004_{24} = \sqrt{0.003^2 + 0.003^2}$)

fraction of solution delivered = $\dfrac{26.207 \pm 0.003 \text{ g}}{503.505 \pm 0.004_{24} \text{ g}}$

= $\dfrac{26.207 \pm 0.01_{14}\% \text{ g}}{503.505 \pm 0.000\ 08_{43}\%}$ = 0.052 049 $\pm$ 0.01$_{14}$%

mol Ag$^+$ delivered = [0.052 049 ($\pm$ 0.01$_{14}$%)][0.028 680 ($\pm$ 0.06$_{16}$%) mol]

= 0.001 4928 ($\pm$ 0.06$_{26}$%) mol = 0.001 492 8 $\pm$ 0.000 000 9$_4$ mol

(b) mol Ag$^+$ = $\dfrac{4.872 \pm 0.003 \text{ g}}{169.873\ 1 \text{ g/mol}} = \dfrac{4.872 \pm 0.061_6\% \text{ g}}{169.873\ 1 \text{ g/mol}}$

= 0.028 680 ($\pm$ 0.06$_{16}$%) mol

volume of solution = 500.00 $\pm$ 0.20 mL

fraction of solution delivered = $\dfrac{25.00 \pm 0.03 \text{ mL}}{500.00 \pm 0.20 \text{ mL}} = \dfrac{25.00 \pm 0.120\%}{500.00 \pm 0.040\%}$

= 0.050 00 $\pm$ 0.1$_{26}$%

mol Ag$^+$ delivered = [0.050 00 ($\pm$ 0.1$_{26}$%)][0.028 680 ($\pm$ 0.06$_{16}$%) mol]

= 0.001 4340 ($\pm$ 0.1$_{40}$%) mol = 0.001 434$_0$ $\pm$ 0.000 002$_0$ mol

(c) $\dfrac{\text{volumetric uncertainty}}{\text{gravimetric uncertainty}} = \dfrac{0.1_{40}\%}{0.06_{26}\%} = 2.24$

In the gravimetric procedure, the largest relative uncertainty is in the mass of AgNO$_3$ that was weighed out: 0.06%.

In the volumetric procedure, the largest relative uncertainty is in the volume of AgNO$_3$ delivered by pipet: 0.12%.

6-12. The amount of standard Na$_2$C$_2$O$_4$ added to K$_2$S$_2$O$_8$ was

mol Na$_2$C$_2$O$_4$ = (0.050 00 L)(0.050 06 mol/L) = 0.002 503 mol

Good Titrations

Some reacted with $K_2S_2O_8$ and some was left over. The amount of $KMnO_4$ required to consume the excess $Na_2C_2O_4$ was

$$\text{mol } KMnO_4 = (0.016\ 52\ L)(0.020\ 13\ mol/L) = 0.000\ 332\ 5\ mol$$

Reaction 6-1 requires 5 mol of $Na_2C_2O_4$ for 2 mol $KMnO_4$. So,

$$\text{mol } Na_2C_2O_4 \text{ in Reaction 6-1} = (\text{mol } MnO_4^-)\left(\frac{5\ \text{mol }Na_2C_2O_4}{2\ \text{mol }MnO_4^-}\right)$$

$$= (0.000\ 332\ 5\ \text{mol }MnO_4^-)\left(\frac{5\ \text{mol }Na_2C_2O_4}{2\ \text{mol }MnO_4^-}\right) = 0.000\ 831\ 4\ \text{mol}$$

The amount of $Na_2C_2O_4$ consumed by reaction with $K_2S_2O_8$ must have been

mol $Na_2C_2O_4$ reacting with $K_2S_2O_8$
= total mol $Na_2C_2O_4$ − mol $Na_2C_2O_4$ reacting with $KMnO_4$
= 0.002 503 mol − 0.000 831 4 mol = 0.001 672 mol $Na_2C_2O_4$

In Reaction 6-4, 1 mol $K_2S_2O_8$ reacts with 1 mol $Na_2C_2O_4$, so there must have been 0.001 672 mol $K_2S_2O_8$ in the impure reagent.

$$\text{mass of } K_2S_2O_8 \text{ in reagent} = (0.001\ 672\ \text{mol})(270.32\ \text{g/mol}) = 0.451\ 9\ \text{g}$$

$$\text{wt\% } K_2S_2O_8 = \frac{0.451\ 9\ \text{g }K_2S_2O_8}{0.507\ 3\ \text{g reagent}} \times 100 = 89.07\ \text{wt\%}$$

6-13. $3MnO_4^- + 5Mo^{3+} + 4H^+ \rightarrow 3Mn^{2+} + 5MoO_2^{2+} + 2H_2O$

(22.11 − 0.07) = 22.04 mL of 0.012 34 M $KMnO_4$ = 0.272 0 mmol of MnO_4^-

This quantity of MnO_4^- will react with

$$\left(\frac{5\ \text{mmol }Mo^{3+}}{3\ \text{mmol }MnO_4^-}\right)(0.272\ 0\ \text{mmol }MnO_4^-) = 0.453\ 3\ \text{mmol of }Mo^{3+}$$

$[Mo^{3+}] = 0.453\ 3\ \text{mmol}/50.00\ \text{mL} = 9.066\ \text{mM}$

6-14. $2MnO_4^- + 5H_2C_2O_4 + 6H^+ \rightarrow 2Mn^{2+} + 10CO_2 + 8H_2O$

12.34 mL of 0.004 321 M $KMnO_4$ = 0.053 32 mmol of MnO_4^-

This quantity of MnO_4^- reacts with $\left(\dfrac{5\ \text{mmol }H_2C_2O_4}{2\ \text{mmol }MnO_4^-}\right)(0.053\ 32\ \text{mmol }MnO_4^-)$

= 0.133 3 mmol of $H_2C_2O_4$. This quantity of $H_2C_2O_4$ is derived from

$\left(\dfrac{2\ \text{mmol }La^{3+}}{3\ \text{mmol }H_2C_2O_4}\right)(0.133\ 3\ \text{mmol }H_2C_2O_4) = 0.088\ 87\ \text{mmol }La^{3+}$.

$[La^{3+}] = 0.088\ 87\ \text{mmol}/25.00\ \text{mL} = 3.555\ \text{mM}$

6-15.

$$50.0 \text{ mL of } 0.089\,9 \text{ M Ce}^{4+} \quad = 4.495 \text{ mmol}$$
$$-10.05 \text{ mL of } 0.043\,7 \text{ M Fe}^{2+} \quad = 0.439 \text{ mmol}$$
$$\overline{\text{Ce}^{4+} \text{ reacting with glycerol} \quad = 4.056 \text{ mmol}}$$

One mole of glycerol requires eight moles of Ce^{4+}:

$$\left(\frac{1 \text{ mmol glycerol}}{8 \text{ mmol Ce}^{4+}}\right)(4.056 \text{ mmol Ce}^{4+}) = 0.507_0 \text{ mmol glycerol}$$

$$= 46.7 \text{ mg glycerol}$$

$$\Rightarrow \text{ original solution} = \frac{46.7 \text{ mg glycerol}}{153.2 \text{ mg solution}} \times 100 = 30.5 \text{ wt\% glycerol}$$

6-16. AgI has the smaller solubility product, so I^- precipitates before SCN^-
If 12.6 mL of Ag^+ are required to precipitate I^-, then $(27.7 - 12.6) = 15.1$ mL are required to precipitate SCN^-.

$$[SCN^-] = \frac{\text{moles of Ag}^+ \text{ needed to react with SCN}^-}{\text{original volume of SCN}^-}$$

$$= \frac{[27.7 \,(\pm 0.3) - 12.6 \,(\pm 0.4) \text{ mL}][0.068\,3 \,(\pm 0.000\,1) \text{ M}]}{50.00 \,(\pm 0.05) \text{ mL}} \quad (a)$$

$$= \frac{[15.1 \,(\pm 0.5)][0.068\,3 \,(\pm 0.000\,1)]}{50.00 \,(\pm 0.05)}$$

$$= \frac{[15.1 \,(\pm 3.31\%)][0.068\,3 \,(\pm 0.146\%)]}{50.00 \,(\pm 0.100\%)} = 0.020\,6 \,(\pm 0.000\,7) \text{ M}$$

In expression (a), I chose to keep the volumes in mL rather than L. As long as the units are consistent and cancel out, this practice is allowed. You could have written all volumes in L.

6-17. (a) $[Cu^+][Br^-] = K_{sp} \Rightarrow x^2 = 5 \times 10^{-9} \Rightarrow [Cu^+] = 7._1 \times 10^{-5}$ M
　　　　xx

(b) $(143.45 \text{ g/mol})(7.1 \times 10^{-5} \text{ mol/L})(0.100 \text{ L}/100 \text{ mL})$
$= 1._0 \times 10^{-3}$ g/100 mL

6-18. (a) $Ag_2CrO_4(s) \overset{K_{sp}}{\rightleftharpoons} 2Ag^+ + CrO_4^{2-}$
　　　FM 331.73　　　　　$2x$　　x

$(2x)^2(x) = 1.2 \times 10^{-12} \Rightarrow x = 6.6_9 \times 10^{-5}$ M

(b) $[Ag^+] = 13.4 \times 10^{-5}$ M
$= 0.014\,4$ g/L $= 0.014\,4$ mg/mL $= 14.4$ µg/mL $= 14.4$ ppm

Good Titrations   45

6-19. $[Ag^+]^4[Fe(CN)_6^{4-}] = K_{sp}$

$[1.0 \times 10^{-6}]^4[Fe(CN)_6^{4-}] = 8.5 \times 10^{-45} \Rightarrow [Fe(CN)_6^{4-}] = 8.5 \times 10^{-21}$ M = 8.5 zM

The answer would not change if other species were present. The equilibrium constant K_{sp} describes the reaction $Ag_4Fe(CN)_6 (s) \rightleftharpoons 4Ag^+ + Fe(CN)_6^{4-}$, regardless of what other chemistry occurs in the solution.

6-20. $AgX(s) \rightleftharpoons Ag^+ + X^-$

FM = 143.32 (X=Cl) $K_{sp} = 1.8 \times 10^{-10}$ for AgCl
FM = 187.77 (X=Br) $K_{sp} = 5.0 \times 10^{-13}$ for AgBr
FM = 234.77 (X=I) $K_{sp} = 8.3 \times 10^{-17}$ for AgI

$[Ag^+][X^-] = [Ag^+]^2 = K_{sp} \Rightarrow [Ag^+] = \sqrt{K_{sp}}$
$[Ag^+] = 1.3_4 \times 10^{-5}$ M for AgCl = 1.4×10^{-3} g/L

$\left(1.4 \times 10^{-3} \frac{g}{L}\right)\left(\frac{10^9 \text{ ng}}{g}\right)\left(\frac{10^{-3} \text{ L}}{mL}\right) = 1\,400$ ppb

$[Ag^+] = 7.0_7 \times 10^{-7}$ M for AgBr = 7.6×10^{-5} g/L = $\boxed{76 \text{ ppb}}$ (best choice)

$[Ag^+] = 9.1_1 \times 10^{-9}$ M for AgI = 9.8×10^{-7} g/L = 0.98 ppb

6-21.   (a) $Hg_2(IO_3)_2 \rightleftharpoons Hg_2^{2+} + 2IO_3^-$ $K_{sp} = 1.3 \times 10^{-18}$
                  FM 750.99 x $2x$

$x(2x)^2 = 1.3 \times 10^{-18} \Rightarrow x = 6.8_8 \times 10^{-7}$ M

$[Hg_2^{2+}] = x = 6.8_8 \times 10^{-7}$ M $[IO_3^-] = 2x = 1.3_8 \times 10^{-6}$ M

(b) $[Hg_2^{2+}](0.010)^2 = 1.3 \times 10^{-18} \Rightarrow [Hg_2^{2+}] = 1.3 \times 10^{-14}$ M

6-22.   Our strategy will be to find the concentration of Ag^+ in equilibrium with 0.10 M anion. The anion requiring the lowest concentration of Ag^+ precipitates first.

Salt	K_{sp}	[Ag$^+$] (M, in equilibrium with 0.10 M anion)		
AgCl	1.8×10^{-10}	$K_{sp}/0.10$	=	1.8×10^{-9}
AgBr	5.0×10^{-13}	$K_{sp}/0.10$	=	5.0×10^{-12}
AgI	8.3×10^{-17}	$K_{sp}/0.10$	=	8.3×10^{-16}
Ag$_2$CrO$_4$	1.2×10^{-12}	$\sqrt{K_{sp}/0.10}$	=	3.5×10^{-6}

The concentration of Ag^+ in equilibrium with 0.10 M I^- is lowest. Therefore, AgI will precipitate before the other silver salts. Following this logic, the order of precipitation is I^- before Br^- before Cl^- before CrO_4^{2-}.

6-23. In step 1, the volume of KSCN required to react with 25.00 mL of 0.102 6 M AgNO$_3$ is 24.22 − 0.02 = 24.20 mL, and the stoichiometry of the reaction is 1:1 (Ag$^+$ + SCN$^-$ → AgSCN). The concentration of KSCN is therefore

$$[\text{KSCN}] = \frac{\text{mol SCN}^-}{\text{volume SCN}^-} = \frac{\text{mol Ag}^+}{\text{volume SCN}^-}$$

$$= \frac{(0.102\ 6\ \text{M})(25.00\ \text{mL})}{24.20\ \text{mL}} = 0.105\ 9_9\ \text{M}$$

In step 2, 20.00 mL of dilute HCl was treated with 25.00 mL of 0.102 6 M AgNO$_3$, which contains (0.025 00 L)(0.102 6 M) = 2.565$_0$ × 10^{-3} mol Ag$^+$. The excess Ag$^+$ required 2.43 − 0.02 = 2.41 mL of 0.105 9$_9$ M SCN$^-$ = 2.55 × 10^{-4} mol SCN$^-$. The quantity of Ag$^+$ that reacted with Cl$^-$ was therefore 2.565$_0$ × 10^{-3} mol − 2.55 × 10^{-4} mol = 2.310$_0$ × 10^{-3} mol. This much Cl$^-$ was present in 20.00 mL of dilute HCl, so the concentration of dilute HCl was (2.310$_0$ × 10^{-3} mol)/(20.00 mL) = 0.115 5$_0$ M. The dilute HCl was prepared by diluting 10.00 mL up to 1.000 L—a 1:100 dilution. Therefore, the original concentrated HCl had a concentration of 100 × 0.115 5$_0$ M = 11.55 M.

6-24. Your spreadsheets should look like those in the book. Answer: x = 0.000 945 4.

6-25. (a) Titration reaction: SO$_4^{2-}$ [from soil] + Ba^{2+} [from BaCl$_2$ (s)] → BaSO$_4$(s)

(b) Prior to adding BaCl$_2$, the 25-mL solution contained 0.000 19 M Cl$^-$. At the end point, the solution contained 0.009 6 M Cl$^-$ from addition of BaCl$_2$.
[Cl$^-$] added from BaCl$_2$ = 0.009 6 − 0.000 19 M = 0.009 4$_1$ M
mmol Cl$^-$ = (0.009 4$_1$ M)(25 mL) = 0.23$_5$ mmol
BaCl$_2$ contains 1 mol Ba^{2+} for every 2 mol Cl$^-$, so
mmol Ba^{2+} added = $\frac{1}{2}$(0.23$_5$ mmol) = 0.11$_8$ mmol

(c) One mole of Ba^{2+} consumes one mole of SO$_4^{2-}$ in the titration. Therefore, 0.11$_8$ mmol SO$_4^{2-}$ must have been present in the 25-mL aqueous extract.

(d) Mass of SO$_4^{2-}$ in extract = (0.11$_8$ mmol SO$_4^{2-}$)(96.06 mg/mmol) = 11.3 mg
wt% SO$_4^{2-}$ in soil = 100 × (11.3 mg SO$_4^{2-}$)/(1 000 mg soil) = 1.1 wt%

CHAPTER 7
GRAVIMETRIC AND COMBUSTION ANALYSIS

7-1. (a) Occluded impurity is trapped inside pockets in the precipitate. Included impurities substitute into the crystal lattice of the precipitate. Adsorbed impurity is bound to the surface of the precipitate.

(b) The first $BaSO_4$ was apparently precipitated in the presence of a great deal of nitrate and a large quantity of nitrate became occluded in the precipitate. When the $BaSO_4$ was redissolved, the only nitrate in solution was already present in the first precipitate, which was much less than the initial concentration of nitrate in solution during the first precipitation. Because there is less nitrate in solution during the second precipitation, less nitrate impurity ends up in the second $BaSO_4$ precipitate. The second precipitate is dissolved and precipitated again to give an even purer produce, because there is even less nitrate in solution during the third precipitation than during the second precipitation.

7-2. One mole of ethoxyl groups produces one mole of AgI.
29.03 mg of AgI = 0.123 65 mmol AgI
The amount of compound analyzed is 25.42 mg/(417 g/mol) = 0.060 96 mmol.
The number of ethoxyl groups per molecule is
$$\frac{0.123\ 65\ \text{mmol ethoxyl groups}}{0.060\ 96\ \text{mmol compound}} = 2.03\ (\approx 2)$$

7-3. Grams of piperazine in sample =
(0.712 9 g of piperazine/g of sample) × (0.050 02 g of sample) = 0.035 66 g
$$\frac{0.035\ 66\ \text{g}}{86.136\ \text{g/mol}} = 4.140 \times 10^{-4}\ \text{mol piperazine, which makes}$$
4.140×10^{-4} mol product = 0.085 38 g

7-4. $$\frac{2.500\ \text{g bis(dimethylglyoximate)nickel(II)}}{288.91\ \text{g/mol}} = 8.653\ 2 \times 10^{-3}\ \text{mol Ni}$$
$$= 0.507\ 85\ \text{g Ni}$$
$$\frac{0.507\ 85\ \text{g Ni}}{1.000\ \text{g sample}} = 50.79\ \text{wt\% Ni}.$$

7-5. 2.07% of 0.998 4 g = 0.020 67 g Ni = 3.521×10^{-4} mol Ni
This quantity requires $(2)(3.521 \times 10^{-4})$ mol of dimethylglyoxime (DMG)
= 0.081 78 g.

A 50.0% excess is (1.5)(0.081 75 g) = 0.122 7 g. The mass of solution containing 0.122 7 g is 0.122 7 g DMG / (0.021 5 g DMG/g solution) = 5.705 g of solution. The volume of solution is 5.704 g/(0.790 g/mL) = 7.22 mL.

7-6. Mol K^+ in unknown = mol $K^+B(C_6H_5)_4^-$ = $\dfrac{1.003 \text{ g}}{358.33 \text{ g/mol}}$

= 2.799 1 mmol = 0.109 44 g K

wt% K = $\dfrac{0.109\ 44 \text{ g K}}{1.263 \text{ g unknown}}$ × 100 = 8.665 wt% K

7-7. Moles of Fe in product (Fe_2O_3) = moles of Fe in sample

$\left(\dfrac{2 \text{ mol Fe}}{\text{mol } Fe_2O_3}\right)\left(\dfrac{0.264 \text{ g } Fe_2O_3}{159.69 \text{ g } Fe_2O_3/\text{mol } Fe_2O_3}\right)$ = 3.306 × 10^{-3} mol of Fe

This many moles of Fe equals 0.919 g of $FeSO_4 \cdot 7H_2O$. Because we analyzed just 2.998 g out of 22.131 g of tablets, the $FeSO_4 \cdot 7H_2O$ in the 22.131 g sample is greater by a factor of (22.131/2.998).

$\left(\dfrac{22.131 \text{ g}}{2.998 \text{ g}}\right)(0.919 \text{ g})$ = 6.784 g

This quantity is the $FeSO_4 \cdot 7H_2O$ content of 20 tablets.

The content in one tablet is (6.784 g)/20 = 0.339 g.

7-8. (a) 70 kg $\left(\dfrac{6.3 \text{ g P}}{\text{kg}}\right)$ = 441 g P in 8.00 × 10^3 L Each milliliter contains

$\dfrac{441 \text{ g P}}{8.00 \times 10^3 \text{ L}}$ = 0.055 1 g/L or 5.5$_1$ mg/100 mL

(b) Fraction of P in one formula mass is

$\dfrac{\left(\dfrac{2 \text{ mol P}}{\text{mol } P_2O_5 \cdot 24MoO_3}\right)\left(\dfrac{30.974 \text{ g P}}{\text{mol P}}\right)}{\dfrac{3\ 596.46 \text{ g } P_2O_5 \cdot 24MoO_3}{\text{mol } P_2O_5 \cdot 24MoO_3}}$ × 100 = 1.722 wt%

P in 0.338 7 g of $P_2O_5 \cdot 24 MoO_3$

 = (0.017 22 g P/g $P_2O_5 \cdot 24 MoO_3$)(0.338 7 g $P_2O_5 \cdot 24 MoO_3$)

 = 5.834 mg P, an amount expected from a dissolved man.

7-9. Formula and atomic masses: Ba (137.327), Cl (35.453), K (39.098), H_2O (18.015), KCl (74.551), $BaCl_2 \cdot 2H_2O$ (244.26)

H_2O lost = 1.783 9 − 1.562 3 = 0.221 6 g = 1.230 1 × 10^{-2} mol of H_2O.

For every two moles of H_2O lost, one mole of $BaCl_2 \cdot 2H_2O$ must have been present. 1.230 1 × 10^{-2} mol of H_2O implies that 6.150 4 × 10^{-3} mol of

Gravimetric and Combustion Analysis   49

BaCl$_2$·2H$_2$O must have been present. This much BaCl$_2$·2H$_2$O equals 1.502 3 g. The Ba and Cl contents of the BaCl$_2$·2H$_2$O are

$$Ba = \left(\frac{137.33 \text{ g Ba}}{244.26 \text{ g BaCl}_2 \cdot 2H_2O}\right)(1.502\,3 \text{ g BaCl}_2 \cdot 2H_2O) = 0.844\,64 \text{ g}$$

$$Cl = \left(\frac{(2)\left(\frac{\text{mol Cl}}{\text{mol BaCl}_2 \cdot 2H_2O}\right)(35.453 \text{ g Cl})}{244.26 \text{ g BaCl}_2 \cdot 2H_2O}\right)(1.502\,3 \text{ g BaCl}_2 \cdot 2H_2O)$$

$$= 0.436\,10 \text{ g}$$

Because the total sample weighs 1.783 9 g and contains 1.502 3 g of BaCl$_2$·2H$_2$O, the sample must contain 1.783 9 − 1.502 3 = 0.281 6 g of KCl, which contains

$$K = \left(\frac{39.098 \text{ g K}}{74.551 \text{ g KCl}}\right)(0.281\,6 \text{ g KCl}) = 0.147\,68 \text{ g}$$

$$Cl = \left(\frac{35.453 \text{ g Cl}}{74.551 \text{ g KCl}}\right)(0.281\,6 \text{ g KCl}) = 0.133\,92 \text{ g}$$

Weight percent of each element:

$$Ba = \frac{0.844\,64}{1.783\,9} \times 100 = 47.35 \text{ wt\%} \qquad K = \frac{0.147\,68}{1.783\,9} \times 100 = 8.279 \text{ wt\%}$$

$$Cl = \frac{0.436\,10 + 0.133\,92}{1.783\,9} \times 100 = 31.95 \text{ wt\%}$$

7-10. (a) Mass of product (CaCO$_3$) = 18.546 7 g − 18.231 1 g = 0.315 6 g

$$\text{mol CaCO}_3 = \left(\frac{0.315\,6 \text{ g}}{100.087 \text{ g/mol}}\right) = 3.153 \times 10^{-3} \text{ mol}$$

The product contains 3.153 mmol Ca = (3.153 × 10^{-3} mol)(40.078 g/mol)
$$= 0.126\,4 \text{ g Ca}$$

$$\text{wt\% Ca} = \frac{0.1264 \text{ g Ca}}{0.632\,4 \text{ g mineral}} \times 100 = 19.98 \text{ wt\% Ca}$$

(b) The solutions are heated before mixing to increase the solubility of the product that will precipitate. If the solution is less supersaturated during the precipitation, crystals form more slowly and grow to be larger and purer than if they precipitate rapidly. The larger crystals are easier to filter.

(c) (NH$_4$)$_2$C$_2$O$_4$ provides oxalate ion to prevent CaC$_2$O$_4$ from redissolving. Also, the ammonium and oxalate ions provide an ionic atmosphere that prevents the precipitate from peptizing (breaking into colloidal particles).

(d) AgNO$_3$ solution is added to the filtrate to test for Cl$^-$ in the filtrate. If Cl$^-$ is present, AgCl(s) will precipitate when Ag$^+$ is added. The source of Cl$^-$ is the HCl used to dissolve the mineral. All of the original solution needs to be washed away, so that no extra material is present that would increase the mass of final product, which should be pure CaCO$_3$(s).

7-11. $C_6H_5CO_2H \ + \ \frac{15}{2}O_2 \ \rightarrow \ 7CO_2 \ + \ 3H_2O$
FM 122.121   FM 44.010   FM 18.015

One mole of C$_6$H$_5$CO$_2$H gives 7 moles of CO$_2$ and 3 moles of H$_2$O
4.635 mg of benzoic acid = 0.037 95 mmol, which gives 0.265 7 mmol CO$_2$
(= 11.69 mg CO$_2$) and 0.113 9 mmol H$_2$O (= 2.051 mg H$_2$O)

7-12. (a) mmol C = mmol CO$_2$ = $\frac{16.432 \text{ mg}}{44.010 \text{ mg/mmol}}$ = 0.373 37 mmol = 4.484 5 mg C

$\frac{4.484 \text{ 5 mg C}}{8.732 \text{ mg unknown}} \times 100 = 51.36$ wt% C

mmol H = 2 × mmol H$_2$O = $\frac{2 \times 2.840 \text{ mg}}{18.015 \text{ mg/mmol}}$ = 0.315 29 mmol

= 0.317 80 mg H = 3.639 wt% H

(b) Atomic ratio (C:H) = 0.373 37 / 0.315 29 = 1.184
The smallest reasonable integer ratio is 6:5 (= 1.2), or C$_6$H$_5$.

7-13. 100 g of compound contain 46.21 g C, 9.02 g H, 13.74 g N, and 31.03 g O. The atomic ratios are

C:H:N:O = $\frac{46.21 \text{ g}}{12.011 \text{ g/mol}} : \frac{9.02 \text{ g}}{1.008 \text{ g/mol}} : \frac{13.74 \text{ g}}{14.007 \text{ g/mol}} : \frac{31.03 \text{ g}}{15.999 \text{ g/mol}}$

= 3.847 : 8.94$_8$: 0.980 9 : 1.939

Dividing by the smallest factor (0.980 9) gives the ratio
C:H:N:O = 3.922 : 9.12 : 1 : 1.977
The empirical formula is probably C$_4$H$_9$NO$_2$.

7-14. 2.378 × 10^{-3} g CO$_2$ / (44.010 g/mol) = 5.403 3 × 10^{-5} mol CO$_2$
= 5.403 3 × 10^{-5} mol C = 6.490 0 × 10^{-4} g C

ppm C = 10^6 × $\frac{6.490 \text{ 0} \times 10^{-4} \text{ g C}}{6.234 \text{ g sample}}$ = 104.1 ppm C

7-15. (a) mmol nitron nitrate product = $\frac{513.6 \text{ mg}}{375.39 \text{ mg/mmol}}$ = 1.368 mmol

Gravimetric and Combustion Analysis

$$[NO_3^-] \text{ in unknown} = \frac{1.368 \text{ mmol}}{50.00 \text{ mL}} = 0.027\,36 \text{ M}$$

(b) Loss of product by dissolution in the wash liquid leads to a systematically low result.

7-16. The mass loss of $18.371 - 17.462 = 0.909$ mg is from $CuO(s) \rightarrow Cu(s)$. The mmol of oxygen lost are $(0.909 \text{ mg})/(15.999\,4 \text{ mg O/mmol}) = 0.056\,8_1$ mmol. Therefore, there must have been $0.056\,8_1$ mmol $CuO = (0.056\,8_1 \text{ mmol})(79.545 \text{ mg CuO/mmol}) = 4.51_9$ mg CuO in the original mixture. The mass of Al_2O_3 in the original mixture must have been $18.371 - 4.51_9 = 13.85_2$ mg. The weight percent of Al_2O_3 is $100 \times (13.85_2 \text{ mg } Al_2O_3)/18.371 \text{ mg sample}) = 75.40$ wt%

7-17. The atomic ratio H:C is

$$\frac{\left(\frac{6.76 \pm 0.12 \text{ g H}}{1.008 \text{ g H/mol}}\right)}{\left(\frac{71.17 \pm 0.41 \text{ g C}}{12.011 \text{ g C/mol}}\right)} = \frac{6.706 \pm 0.119}{5.925 \pm 0.034\,1} = \frac{6.706 \pm 1.78\%}{5.925 \pm 0.576\%} = 1.132 \pm 0.021$$

If we define the stoichiometry coefficient for C to be 8, then the stoichiometry coefficient for H is $8(1.132 \pm 0.021) = 9.06 \pm 0.17$.

The atomic ratio N:C is $\dfrac{\left(\dfrac{10.34 \pm 0.08 \text{ g N}}{14.007 \text{ g N/mol}}\right)}{\left(\dfrac{71.17 \pm 0.41 \text{ g C}}{12.011 \text{ g C/mol}}\right)} = \dfrac{0.738\,2 \pm 0.005\,7}{5.925 \pm 0.034\,1}$

$$= \frac{0.738\,2 \pm 0.774\%}{5.925 \pm 0.576\%} = 0.124\,6 \pm 0.001\,2$$

If we define the stoichiometry coefficient for C to be 8, then the stoichiometry coefficient for N is $8(0.124\,6 \pm 0.001\,2) = 0.996\,8 \pm 0.009\,6$.

The empirical formula is reasonably expressed as $C_8H_{9.06\pm0.17}N_{0.997\pm0.010}$

7-18. The reaction between H_2SO_4 and NaOH can be written

$$H_2SO_4 + 2NaOH \rightarrow 2H_2O + Na_2SO_4$$

One mole of H_2SO_4 requires two moles of NaOH. In 3.01 mL of 0.015 76 M NaOH, there are $(0.003\,01 \text{ L})(0.015\,76 \text{ mol/L}) = 4.74_4 \times 10^{-5}$ mol of NaOH. The moles of H_2SO_4 must have been $\left(\frac{1}{2}\right)(4.74_4 \times 10^{-5}) = 2.37_2 \times 10^{-5}$ mol. Because one mole of H_2SO_4 contains one mole of S, there must have been $2.37_2 \times 10^{-5}$ mol of S ($= 0.760_6$ mg). The percentage of S in the sample is

$$\frac{0.760_6 \text{ mg S}}{6.123 \text{ mg sample}} \times 100 = 12.4 \text{ wt\% S}$$

7-19. (a) $BaSO_4$ FM = 233.390 g/mol;   1 mol of BaSO4 contains 1 mol S

$$\text{mol } BaSO_4 = \frac{0.352\ 9 \text{ g } BaSO_4}{233.390 \text{ g/mol}} = 1.512_1 \text{ mmol} = \text{mol S}$$

mass of S in $BaSO_4$ = mass of S in coal
= $(1.512_1 \text{ mmol S})(32.065 \text{ g/mol}) = 0.048\ 48_2$ g S

$$\text{wt\% S in coal} = 100 \times \frac{0.048\ 48_2 \text{ g S}}{2.136 \text{ g of coal}} = 2.270 \text{ wt\%}$$

(b) SO_2 FM = 64.064 g/mol;   S atomic mass = 32.065 g/mol

1 mol S produces 1 mol SO_2; 32.065 g S produces 64.064 g SO_2

1 g S produces (64.064 g SO_2/32.065 g S)(1 g S) = 1.997 9 g SO_2

SO_2 produced by 8 million tons of coal containing 2 wt% S

$$= \left(0.02 \frac{\text{ton S}}{\text{ton coal}}\right)(8 \text{ million tons coal})\left(1.997\ 9 \frac{\text{ton } SO_2}{\text{ton S}}\right)$$

= 0.32 million tons SO_2

(c) The coal produces 7 wt% ash. 8 million tons of coal will produce
(0.07)(8 million tons) = 0.56 million tons of ash

7-20. (a) In the case of Fe^{3+} (atomic mass = 55.845), the initial solution contained 10.0 μg = 0.179_1 μmol. The final solution volume was 10.00 mL and the observed concentration was 17.6 μM. There are $(17.6 \times 10^{-6} \text{ M})(0.010\ 00 \text{ L})$ = 1.76×10^{-7} mol Fe^{3+} = 9.83 μg Fe^{3+} in the final solution. The recovery is (9.83 μg/10.0 μg) × 100 = 98.3%. Here are the other results:

Element	Atomic mass	Found (μM)	Recovery (%)
Fe^{3+}	55.845	17.6	98.3
Pb^{2+}	207.2	5.02	104.0
Cd^{2+}	112.411	8.77	98.6
In^{3+}	114.818	8.50	97.6
Cr^{3+}	51.996	< 0.05	<0.3
Mn^{2+}	54.938	6.64	36.5
Co^{2+}	58.933	1.09	6.4
Ni^{2+}	58.693	< 0.05	<0.3
Cu^{2+}	63.546	6.96	4.2

(b) It appears that Fe^{3+}, Pb^{2+}, Cd^{2+}, and In^{3+} are gathered quantitatively.

(c) The initial sample solution has a volume of 100.0 mL and the final solution has a volume of 10.00 mL, ideally containing the same quantity of analyte. Therefore, the analyte is concentrated by a factor of 10.0 because the same number of moles are present in one-tenth of the initial volume.

7-21. For the 14 manual method results, the mean is 2.926_3 and the standard deviation is 0.041_6. For the 21 automated results, the mean is 2.990_6 and the standard deviation is 0.032_5. We will compare the means with the t test. The pooled standard deviation is

$$s_{pooled} = \sqrt{\frac{s_1^2(n_1-1) + s_2^2(n_2-1)}{n_1+n_2-2}}$$

$$= \sqrt{\frac{0.0416^2(14-1) + 0.0325^2(21-1)}{14+21-2}} = 0.03636$$

$$t = \frac{|\bar{x}_1 - \bar{x}_2|}{s_{pooled}}\sqrt{\frac{n_1 n_2}{n_1+n_2}} = \frac{|2.9263 - 2.9906|}{0.03636}\sqrt{\frac{14 \cdot 21}{14+21}} = 5.1_3$$

For $14 + 21 - 2 = 33$ degrees of freedom, the tabulated value of t is approximately 2.04 at the 95% confidence level and 2.75 at the 99% confidence level. Because $t_{calculated} > t_{table}$, the difference is significant. The two methods give statistically different results.

CHAPTER 8
INTRODUCING ACIDS AND BASES

8-1. (a) acid-base pairs: HCN/CN^- HCO_2H/HCO_2^-

(b) H_2O/OH^- HPO_4^{2-}/PO_4^{3-}

(c) H_2O/OH^- HSO_3^-/SO_3^{2-}

8-2. $[H^+] > [OH^-]$, $[OH^-] > [H^+]$

8-3. (a) $pH = -\log(10^{-4}) = 4$

(b) $[H^+] = K_w/[OH^-] = 10^{-14}/10^{-5} = 10^{-9}$ M $\Rightarrow pH = -\log(10^{-9}) = 9$

(c) $pH = -\log(5.8 \times 10^{-4}) = 3.24$

(d) $[H^+] = K_w/[OH^-] = K_w/(5.8 \times 10^{-5}) = 1.7_2 \times 10^{-10}$ M $\Rightarrow pH = 9.76$

8-4. (a) $\left(\dfrac{10^{-6} \text{ M}}{1 \text{ μM}}\right)(0.035 \text{ μM}) = 3.5 \times 10^{-8}$ M

$pH = -\log(3.5 \times 10^{-8}) = 7.46$

(b) $[OH^-] = K_w/[H^+] = K_w/(3.5 \times 10^{-8}) = 2.86 \times 10^{-7}$ M

8-5. $[H^+] = 10^{-pH} = 10^{-4.6} = 2.5 \times 10^{-5}$ M

One mole of H_2SO_4 provides one mole of H^+, so the concentration of H_2SO_4 needed to give 2.5×10^{-5} M H^+ is 2.5×10^{-5} M.

8-6. $[La^{3+}][OH^-]^3 = K_{sp} = 2 \times 10^{-21}$

$[OH^-]^3 = K_{sp}/[La^{3+}] = (2 \times 10^{-21})/(0.010)$

$\Rightarrow [OH^-] = 5._8 \times 10^{-7}$ M $\Rightarrow pH = 7.8$

8-7. See Table 8-1.

8-8. Weak acids: RCO_2H $R_3NH^+X^-$ $M(H_2O)_x^{n+}$
                        Carboxylic   Ammonium   Aqueous
                        acids   salts   metal ion
                                                                 with
                                                                 charge $n \geq 2$

Weak bases: R_3N: $RCO_2^-M^+$
               Amines   Carboxylate
                                    salts

8-9. (a) $[H^+] = 0.010$ M $\Rightarrow pH = -\log[H^+] = 2.00$

(b) $[OH^-] = 0.035$ M $\Rightarrow [H^+] = K_w/[OH^-] = 2.8_6 \times 10^{-13}$ M $\Rightarrow pH = 12.54$

(c) $[H^+] = 0.030\ M \Rightarrow pH = 1.52$

(d) $[H^+] = 3.0\ M \Rightarrow pH = -0.48$

(e) $[OH^-] = 0.010\ M \Rightarrow [H^+] = 1.0 \times 10^{-12}\ M \Rightarrow pH = 12.00$

8-10. (a) $Cl_3CCO_2H \rightleftharpoons Cl_3CCO_2^- + H^+$

$Cu^{2+} + H_2O \rightleftharpoons Cu(OH)^+ + H^+$

$C_6H_5-NH_3^+ \rightleftharpoons C_6H_5-NH_2 + H^+$

(b) K_a for trichloroacetic acid (0.3) is greater than K_a for Cu^{2+} (3×10^{-8}) and for anilinium ion (2.51×10^{-5}), so trichloroacetic acid is the stronger acid.

8-11. (a) $C_6H_5N: + H_2O \rightleftharpoons C_6H_5NH^+ + OH^-$ $K_b = K_w/K_a = 1.6 \times 10^{-9}$

$HOCH_2CH_2S^- + H_2O \rightleftharpoons HOCH_2CH_2SH + OH^-$ $K_b = K_w/K_a = 5.3 \times 10^{-5}$

$CN^- + H_2O \rightleftharpoons HCN + OH^-$ $K_b = K_w/K_a = 1.6 \times 10^{-5}$

(b) Sodium 2-mercaptoethanol has the largest K_b, so it is the strongest base.

8-12. $HO-\underset{O}{\overset{O}{S}}-OH + HO-\underset{O}{\overset{O}{S}}-OH \rightleftharpoons HO-\underset{O}{\overset{^+OH}{S}}-OH + HO-\underset{O}{\overset{O}{S}}-O^-$

(or $2H_2SO_4 \rightleftharpoons H_3SO_4^+ + HSO_4^-$)

8-13. $C_6H_{10}NH + H_2O \rightleftharpoons C_6H_{10}NH_2^+ + OH^-$

$C_6H_5-CO_2^- + H_2O \rightleftharpoons C_6H_5-CO_2H + OH^-$

8-14. $OCl^- + H_2O \rightleftharpoons HOCl + OH^-$ $K_b = K_w/K_a = 3.3 \times 10^{-7}$

8-15. $[OH^-] = 2 \times 3.0 \times 10^{-5} = 6.0 \times 10^{-5}\ M$

$pH = -\log(K_w/[OH^-]) = -\log(1.0 \times 10^{-14}/6.0 \times 10^{-5}) = 9.78$

8-16. $[H^+] = 10^{-pH} = 10^{-11.65} = 2.2 \times 10^{-12}$ M

8-17. HA $\rightleftharpoons$ A$^-$ + H$^+$
Let $x = [H^+] = [A^-]$ and $0.0100 - x = [HA]$.
$$K_a = \frac{[H^+][A^-]}{[HA]} = \frac{(x)(x)}{0.0100 - x} = 1.00 \times 10^{-4}$$
$\Rightarrow x = 9.51 \times 10^{-4}$ M $\Rightarrow$ pH $= -\log x = 3.02$
Fraction of dissociation $= \dfrac{x}{F} = \dfrac{9.51 \times 10^{-4}}{0.0100} = 9.51 \times 10^{-2}$

8-18. C$_6$H$_5$OH $\rightleftharpoons$ C$_6$H$_5$O$^-$ + H$^+$
$0.150 - x$ x x

$\dfrac{x^2}{0.150 - x} = 1.01 \times 10^{-10}$
$\Rightarrow x = 3.89 \times 10^{-6}$

pH $= -\log x = 5.41$
$\dfrac{[A^-]}{[HA] + [A^-]} = \dfrac{3.89 \times 10^{-6}}{0.150} = 2.59 \times 10^{-5}$

8-19. C$_5$H$_5$NH$^+$ $\rightleftharpoons$ C$_5$H$_5$N + H$^+$ $K_a = 6.3 \times 10^{-6}$
$F - x$ x x

$\dfrac{x^2}{0.0850 - x} = K_a \Rightarrow x = 7.29 \times 10^{-4} \Rightarrow$ pH $= 3.14$

$[C_5H_5N] = 7.29 \times 10^{-4}$ M; $[C_5H_5NH^+] = 0.0850 - 7.29 \times 10^{-4} = 0.0843$ M;
$[Br^-] = 0.0850$ M (= formal concentration of pyridinium bromide)

8-20. Zn^{2+} behaves as a weak acid in water and NO$_3^-$ has negligible basic character because it is the conjugate base of the strong acid HNO$_3$.

$\text{Zn}(H_2O)_w^{2+} \rightleftharpoons \text{Zn}(H_2O)_{w-1}(OH)^+ + H^+$ $K_a = 10^{-9.0} = 1.0 \times 10^{-9}$
$F - x$ x x

$\dfrac{x^2}{0.10 - x} = K_a \Rightarrow x = 1.0 \times 10^{-5} \Rightarrow$ pH $= 5.00$

8-21. If the pH is 2.36, then $[H^+] = 10^{-2.36}$. In a solution of HA, $[A^-] = [H^+] = 10^{-2.36}$.
Also, $[HA] = F - [A^-] = 0.100 - 10^{-2.36}$.

       HA $\rightleftharpoons$ H$^+$ + A$^-$
$0.100 - 10^{-2.36}$ $10^{-2.36}$ $10^{-2.36}$

Introducing Acids and Bases

$$K_a = \frac{(10^{-2.36})(10^{-2.36})}{0.100 - 10^{-2.36}} = 1.99 \times 10^{-4} \qquad pK_a = -\log K_a = 3.70$$

8-22. The statement "HA is 0.15% dissociated" means $\dfrac{[A^-]}{[HA] + [A^-]} = 0.0015$.

$$\begin{array}{cccc} HA & \rightleftharpoons & H^+ & + & A^- \\ 0.0222 - x & & x & & x \end{array}$$

$$\frac{[A^-]}{[HA]+[A^-]} = \frac{x}{0.0222 - x + x} = 0.0015 \Rightarrow x = 3.33 \times 10^{-5} \text{ M}$$

$$K_a = \frac{x^2}{0.0222 - x} = \frac{(3.33 \times 10^{-5})^2}{0.0222 - (3.33 \times 10^{-5})} = 5.0 \times 10^{-8}$$

$$pK_a = -\log K_a = 7.30$$

8-23. $\begin{array}{cccc} (CH_3)_3NH^+ & \rightleftharpoons & (CH_3)_3N & + & H^+ \\ F - x & & x & & x \end{array}$ $\qquad K_a = 1.59 \times 10^{-10}$

$$\frac{x^2}{0.060 - x} = K_a \Rightarrow x = 3.0_9 \times 10^{-6} \Rightarrow pH = 5.51$$

$[(CH_3)_3N] = x = 3.1 \times 10^{-6}$ M; $[(CH_3)_3NH^+] = F - x = 0.060$ M

8-24. (a) $\begin{array}{cccc} HA & \rightleftharpoons & H^+ & + & A^- \\ F - x & & x & & x \end{array}$

$$\frac{x^2}{F - x} = K_a = 9.8 \times 10^{-5} \Rightarrow x = 9.4 \times 10^{-4} \Rightarrow pH = 3.03$$

$$\frac{[A^-]}{[HA]+[A^-]} = \frac{x}{F} = 0.094 = 9.4\%$$

(b) In a 10^{-10} M solution of HA, there is too little HA to affect the pH of the water. The pH is essentially the same as that of pure water, which is 7.00. From the K_a equilibrium, we can write

$$[A^-] = \frac{K_a}{[H^+]}[HA] = \frac{9.8 \times 10^{-5}}{1.0 \times 10^{-7}}[HA] = 980[HA]$$

Now substitute 980[HA] for [A$^-$] in the expression for fraction of dissociation

$$\frac{[A^-]}{[HA]+[A^-]} = \frac{980[HA]}{[HA]+980[HA]} = \frac{980}{981} = 0.999 = 99.9\%$$

8-25. The acid dissociation constant for BH^+ is K_w/K_b.

$$BH^+ \overset{K_a}{\rightleftharpoons} B + H^+ \qquad K_a = K_w/K_b = 1.00 \times 10^{-10}$$
$$\underset{0.100-x}{} \quad \underset{x}{} \quad \underset{x}{}$$

$$\frac{x^2}{0.100-x} = 1.00 \times 10^{-10} \Rightarrow x = [B] = [H^+] = 3.16 \times 10^{-6}\ M \Rightarrow pH = 5.50$$

8-26. K_a for cyclohexylammonium ion (listed in the appendix) = 2.71×10^{-11}

K_b for cyclohexylamine = $K_w/K_a = 3.69 \times 10^{-4}$

8-27. (a) $HOCH_2CH_2NH_2 + H_2O \overset{K_b}{\rightleftharpoons} HOCH_2CH_2NH_3^+ + OH^-$

(b) $HOCH_2CH_2NH_3^+ \overset{K_a}{\rightleftharpoons} HOCH_2CH_2NH_2 + H^+$

8-28. $(CH_3)_3N + H_2O \rightleftharpoons (CH_3)_3NH^+ + OH^- \qquad K_b = K_w/K_a = 6.29 \times 10^{-5}$
$\underset{F-x}{} \qquad \underset{x}{} \quad \underset{x}{}$

$$\frac{x^2}{0.060-x} = K_b \Rightarrow x = 1.9_1 \times 10^{-3} \Rightarrow pH = -\log\frac{K_w}{x} = 11.28$$

$[(CH_3)_3NH^+] = x = 1.9_1 \times 10^{-3}\ M;\quad [(CH_3)_3N] = F - x = 0.058\ M$

8-29. $CH_3CO_2^- + H_2O \rightleftharpoons CH_3CO_2H + OH^- \qquad K_b = K_w/K_a = 5.71 \times 10^{-10}$
$\underset{F-x}{} \qquad \underset{x}{} \quad \underset{x}{}$

0.100 M acetate: $\dfrac{x^2}{(1.00 \times 10^{-1})-x} = K_b \Rightarrow x = 7.56 \times 10^{-6} \Rightarrow pH = 8.88$

$$\frac{[HA]}{[HA]+[A^-]} = \frac{x}{F} = 0.007\,56\%$$

0.010 0 M acetate: $\dfrac{x^2}{(1.00 \times 10^{-2})-x} = K_b \Rightarrow x = 2.39 \times 10^{-6} \Rightarrow pH = 8.38$

$$\frac{[HA]}{[HA]+[A^-]} = \frac{x}{F} = 0.023\,9\%$$

For 1.00×10^{-12} M acetate, pH = 7.00 because there is too little acetate to alter the pH of pure water. If pH = 7.00, then $[OH^-] = 10^{-7.00}$ and we can say

$$[HA] = \frac{K_b[A^-]}{[OH^-]} = \frac{5.71 \times 10^{-10}\,[A^-]}{10^{-7.00}} = 5.71 \times 10^{-3}\,[A^-]$$

Now substitute $[HA] = 5.71 \times 10^{-3}\,[A^-]$ into the expression for fraction of association:

$$\frac{[HA]}{[HA]+[A^-]} = \frac{5.71 \times 10^{-3}\,[A^-]}{(5.71 \times 10^{-3} + 1)[A^-]} = 0.568\%$$

Introducing Acids and Bases

8-30. NaCN dissociates into Na^+ (which is neither an acid nor a base) and CN^-, the conjugate base of the weak acid, HCN.

$$CN^- + H_2O \rightleftharpoons HCN + OH^- \quad K_b = K_w/K_a = 1.6 \times 10^{-5}$$
$$F - x \qquad\qquad\quad x \quad\;\; x$$

$$\frac{x^2}{0.050-x} = K_b \Rightarrow x = 8.9 \times 10^{-4} \Rightarrow pH = -\log\frac{K_w}{x} = 10.95$$

8-31. NaOCl dissociates into Na^+ (which is neither an acid nor a base) and OCl^-, the conjugate base of the weak acid, HOCl (hypochlorous acid).

$$OCl^- + H_2O \rightleftharpoons HOCl + OH^-$$
$$0.026 - x \qquad\qquad\;\; x \qquad\; x$$

$$K_b = K_w/K_a = K_w/(3.0 \times 10^{-8}) = 3.33 \times 10^{-7}$$

$$\frac{x^2}{0.026-x} = K_b \Rightarrow x = [OH^-] = 9.3 \times 10^{-5} \text{ M}$$

$$pH = -\log(K_w/[OH^-]) = 9.97$$

$$\frac{[HOCl]}{[HOCl]+[OCl^-]} = \frac{x}{F} = \frac{9.3 \times 10^{-5}}{0.026} = 0.0036$$

8-32. If pH = 10.50, then $[H^+] = 10^{-10.50}$

$[OH^-] = K_w/[H^+] = 10^{-3.50} = 3.16 \times 10^{-4}$ M

$$B + H_2O \rightleftharpoons BH^+ + OH^-$$
$$0.030 - x \qquad\qquad x \qquad\; x$$

$$K_b = \frac{x^2}{0.030-x} = \frac{(3.16 \times 10^{-4})^2}{0.030 - 3.16 \times 10^{-4}} = 3.3_7 \times 10^{-6}$$

8-33. $\alpha_{BH^+} = \dfrac{[BH^+]}{[B]+[BH^+]} = \dfrac{x}{F} = 0.0027$

$F = 0.030$ M $\Rightarrow x = 8.10 \times 10^{-5}$ M.

$$K_b = \frac{x^2}{F-x} = \frac{(8.10 \times 10^{-5})^2}{0.030 - 8.10 \times 10^{-5}} = 2.2 \times 10^{-7}$$

8-34. Amines (RNH_2) are protonated to ammonium salts (RNH_3^+) by the acid in lemon juice. Ammonium salts have lower vapor pressure than amines, so they do not smell as much.

8-35. $Cr(ClO_4)_3$ dissociates to $Cr^{3+} + 3ClO_4^-$. Perchlorate anion has no acid-base chemistry. Cr^{3+} is acidic because of the reaction:

$$Cr(H_2O)_w^{3+} \underset{}{\overset{K_a}{\rightleftharpoons}} Cr(H_2O)_{w-1}(OH)^{2+} + H^+$$
$$0.010-x \qquad\qquad x \qquad\qquad x$$

$$\frac{x^2}{0.010-x} = 10^{3.80} \Rightarrow x = 1.18 \times 10^{-3}\ M$$

$$pH = -\log x = 2.93$$

Fraction of $Cr(H_2O)_{w-1}(OH)^{2+} = \dfrac{[Cr(H_2O)_{w-1}(OH)^{2+}]}{[Cr^{3+}]+[Cr(H_2O)_{w-1}(OH)^{2+}]} = \dfrac{x}{0.010} = 0.118$

8-36. $\dfrac{x^2}{F-x} = K \Rightarrow x^2 + Kx - KF = 0 \Rightarrow x = \dfrac{-K + \sqrt{K^2 + 4KF}}{2}$

The negative root was rejected in the solution to the quadratic equation. Here is a spreadsheet that checks Ask Yourself Problem 8-F (a):

	A	B	C	D	E
1	Solving x^2/(F-x) = K with the quadratic equation				
2					
3	F =	x =			
4	0.1	9.950E-04			
5	K =	B4 = 0.5*(-A6+SQRT(A6^2+4*A6*A4))			
6	0.00001				

8-37.

	A	B
1	Using Excel GOAL SEEK	
2		
3	x	x^2/(F-x)
4	1.0000E-02	1.111E-03
5	F =	
6	0.1	

	A	B
1	Using Excel GOAL SEEK	
2		
3	x	x^2/(F-x)
4	9.9501E-04	1.000E-05
5	F =	
6	0.1	

Initial spreadsheet   After executing Goal Seek

Set up the initial spreadsheet and guess a value of x in cell A4. Compute $x^2/(F-x)$ in cell B4 with the guessed value of x. Use Goal Seek to vary cell A4 until cell B4 is equal to K, which is 1e-5 in this problem. To obtain numerical precision, click the Microsoft Office button at the upper left of the spreadsheet. At the bottom of the Office window, click Excel Options. At the left side of the next window, select Formulas. In Calculation Options, set Maximum Change to 1e-15. Click OK. We just told the computer that cell B4 needs to be precise to 10^{-15}. Select the Data ribbon. In Data Tools, click What-If Analysis and select

Goal Seek. In the resulting window, Set cell <u>B4</u> To value <u>1e-5</u> By changing cell <u>A4</u>. Click OK and Excel finds $x = 9.95\text{e-}4$ in cell A4. If you do not see enough digits in cell A4, drag the separator between columns A and B to the right to expand the cell. To select the number of decimal places displayed, highlight cell A4 and select Number from the Home ribbon.

8-38. The volume of rain in each season is proportional to the rainfall depth in cm, which is reported in the table. The moles of H^+ falling each season are proportional to the volume of rain times the average concentration of H^+ in each season. For the winter, the average concentration of H^+ is $10^{-pH} = 10^{-4.40} = 3.98 \times 10^{-5}$ M. The moles of H^+ falling in the winter are proportional to the product $(3.98 \times 10^{-5} \text{ M})(17.3 \text{ cm})$. In the spreadsheet below, column E is the product of columns D and B. Beneath column E are shown the sum of E4 through E7 and the mean $[H^+]$ computed by dividing the sum in E9 by the total rainfall (80.3 cm). The weighted average mean pH for the year in cell E11 is $-\log(\text{E10}) = 4.65$.

	A	B	C	D	E
1	Weighted average pH in precipitation at Philadelphia, 1990				
2					
3	Season	rainfall (cm)	pH	[H+] (M)	[H+]*rainfall
4	winter	17.3	4.40	3.98E-05	0.0006887
5	spring	30.5	4.68	2.09E-05	0.0006372
6	summer	17.8	4.68	2.09E-05	0.0003719
7	fall	14.7	5.10	7.94E-06	0.0001168
8					
9	total =	80.3		sum =	0.0018146
10				mean [H+] =	2.26E-05
11				mean pH =	4.65
12					
13	D4 = 10^(−C4)				
14	E4 = D4*B4				
15	E9 = SUM(E4:E7)				
16	E10 = E9/SUM(B4:B7)				
17	E11 = −LOG10(E10)				

CHAPTER 9
BUFFERS

9-1. The buffer consists of an acid, HA, and its conjugate base, A⁻. The pH is governed by the Henderson-Hasselbalch equation: $pH = pK_a + \log([A^-]/[HA])$. When H⁺ is added to the mixture, a stoichiometric amount of A⁻ is converted to HA. If we begin with [A⁻]/[HA] = 1:1 and the H⁺ added is enough to eat up 10% of the A⁻, then the new ratio [A⁻]/[HA] is 0.9/1.1 and the pH changes by just $\log(0.9/1.1) = -0.09$ pH units.

9-2. The pH of a buffer depends on the ratio of the concentrations of HA and A⁻ ($pH = pK_a + \log [A^-]/[HA]$). When the volume of solution is changed, both concentrations are affected equally and their ratio does not change.

9-3. If we know the concentration of an acid and its conjugate base, we can find the pH no matter what else is in the solution because all equilibria must be satisfied by the same concentration of H⁺. Putting the quotient [acrylate]/[acrylic acid] = 0.75 into the Henderson-Haselbalch equation for acrylic acid gives

$$pH = 4.25 + \log 0.75 = 4.13$$

9-4. (a) $pH = pK_a + \log \dfrac{[A^-]}{[HA]} = pK_a + \log 100 = pK_a + 2$

(b) $pH = pK_a - 3 = pK_a + \log \dfrac{[A^-]}{[HA]} \Rightarrow \log \dfrac{[A^-]}{[HA]} = -3$

$10^{\log([A^-]/[HA])} = 10^{-3} \Rightarrow \dfrac{[A^-]}{[HA]} = \dfrac{1}{1\,000}$

(c) $pH = pK_a + \log \dfrac{[A^-]}{[HA]} = pK_a + \log 10^{-4} = pK_a - 4$

9-5. The indicator is a weak acid with $pK_a \approx 8$. We know this because pH 8 is the middle of the color transition range in Table 9-3. The acidic form is yellow. The basic form is red. At high pH, the red form predominates and the solution looks red. At low pH, the yellow form predominates and the solution looks yellow. When there is a mixture of both forms in comparable amounts, the solution is expected to be a combination of yellow and red, which is orange. At pH 10, there will be ~100 times as much base as acid because $pH = pK_a + \log([base]/[acid])$ and pH is 2 units above pK_a. Therefore, the solution will appear red. At pH 6, there will be ~100 times as much acid as base, so the solution will look yellow. The

Buffers
63

color transition requires ~2 pH units because the quotient [base]/[acid] changes from 10:1 at pH = pK_a + 1 to 1:10 at pH = pK_a − 1. When one form is 10 times as concentrated as the other form, we see only the color of the dominant form.

9-6. We begin by finding pK_a for the congugate acid, HIO_3, from the relationship $K_a K_b = K_w$, or $K_a = K_w/K_b = 10^{-14.00}/10^{-13.83} = 10^{-0.17}$. $pK_a = -\log K_a = 0.17$. The Henderson-Hasselbalch equation is

$$pH = 0.17 + \log \frac{[IO_3^-]}{[HIO_3]} \Rightarrow \frac{[HIO_3]}{[IO_3^-]} = 10^{(pK_a - pH)}$$

pH	$[HIO_3]/[IO_3^-]$
7.00	1.5×10^{-7}
1.00	0.15

9-7. $K_b = 10^{-10.85}$; $K_a = K_w/K_b \Rightarrow K_a = 10^{-3.15} \Rightarrow pK_a = 3.15$

$$pH = pK_a + \log \frac{[NO_2^-]}{[HNO_2]} = 3.15 + \log \frac{[NO_2^-]}{[HNO_2]}$$

(a) $2.00 = 3.15 + \log \frac{[NO_2^-]}{[HNO_2]} \Rightarrow 10^{-1.15} = \frac{[NO_2^-]}{[HNO_2]}$

$\Rightarrow [HNO_2]/[NO_2^-] = 14$

(b) If pH = 10.00, $[HNO_2]/[NO_2^-] = 1.4 \times 10^{-7}$.

9-8. Substitute the known pH into the Henderson-Hasselbalch equation and solve for the quotient $[CH_3NH_2]/[CH_3NH_3^+]$.

$$pH = 10.632 + \log \frac{[CH_3NH_2]}{[CH_3NH_3^+]} \Rightarrow \frac{[CH_3NH_2]}{[CH_3NH_3^+]} = 10^{(pH - 10.632)}$$

pH	$[CH_3NH_2]/[CH_3NH_3^+]$
4.00	2.33×10^{-7}
10.632	1.00
12.00	23.3

9-9. $pH = pK_a + \log \frac{[A^-]}{[HA]} = 3.46 + \log \frac{(5.13 \text{ g})/(112.13 \text{ g/mol})}{(2.53 \text{ g})/(74.04 \text{ g/mol})} = 3.59$

9-10. (a) $pH = pK_a + \log \frac{[B]}{[BH^+]} = 8.20 + \log \frac{[(1.00 \text{ g})/(74.08 \text{ g/mol})]}{[(1.00 \text{ g})/(110.54 \text{ g/mol})]} = 8.37$

(b) In this case, we know the moles of BH$^+$ and we need to find the moles of B by substitution into the Henderson-Hasselbalch equation.

$$pH = pK_a + \log \frac{\text{mol B}}{\text{mol BH}^+}$$

$$8.00 = 8.20 + \log \frac{\text{mol B}}{[(1.00\text{ g})/(110.54\text{ g/mol})]}$$

$\Rightarrow$ mol B = 0.005 708 = 0.423 g of glycine amide

(c) 5.00 mL of 0.100 M HCl = 0.000 500 mol H$^+$. When we add this much H$^+$ to the existing mixture of B and BH$^+$, the concentrations change as follows:

	B	+	H$^+$	$\rightarrow$	BH$^+$
Initial mol:	0.013 498		0.000 500		0.009 046
Final mol:	0.012 998		—		0.009 546

$$pH = 8.20 + \log\left(\frac{0.012\,998}{0.009\,546}\right) = 8.33$$

(d) 10.00 mL of 0.100 M NaOH = 0.001 00 mol OH$^-$. Adding this much OH$^-$ to the concentrations in part (c) gives

	BH$^+$	+	OH$^-$	$\rightarrow$	B
Initial mol:	0.009 546		0.001 000		0.012 998
Final mol:	0.008 546		—		0.013 998

$$pH = 8.20 + \log\left(\frac{0.013\,998}{0.008\,546}\right) = 8.41$$

9-11. (a) imidazole + H$_2$O $\xrightleftharpoons{K_b}$ imidazolium$^+$ + OH$^-$

imidazolium$^+$ $\xrightleftharpoons{K_a}$ imidazole + H$^+$

(b) $pH = 6.993 + \log \dfrac{1.00\text{ g}/(68.08\text{ g/mol})}{1.00\text{ g}/(104.54\text{ g/mol})} = 7.18$

(c) 2.30 mL of 1.07 M HClO$_4$ = 2.46 mmol H$^+$

Buffers   65

	B	+	H⁺	→	BH⁺
Initial mmol:	14.6₉		2.46		9.57
Final mmol:	12.2₃		—		12.0₃

$$pH = 6.993 + \log \frac{12.2_3}{12.0_3} = 7.00$$

(d) The imidazole must be half neutralized to obtain pH = pK_a. There are 14.6₉ mmol of imidazole, which will require $\frac{1}{2}$(14.6₉ mmol) = 7.34 mmol of HClO₄ = 6.86 mL.

9-12. (a) $pH = pK_a + \log \frac{[A^-]}{[HA]} \Rightarrow pH = 2.865 + \log \frac{0.040\,0}{0.080\,0} = 2.56$

(b) 0.080 mol of HNO₃ + 0.080 mol of Ca(OH)₂ react completely, thus leaving an excess of 0.080 mol of OH⁻. This much OH⁻ converts 0.080 mol of ClCH₂CO₂H into 0.080 mol of ClCH₂CO₂⁻. The final concentrations are
[ClCH₂CO₂⁻] = 0.020 + 0.080 = 0.100 M
[ClCH₂CO₂H] = 0.180 − 0.080 = 0.100 M.
Because [A⁻] = [HA], pH = pK_a = 2.86.

9-13. 213 mL of 0.006 66 M 2,2′-bipyridine = 1.41₉ mmol base. We will add x mol H⁺ to get a pH of 4.19.

	2,2′-bipyridine	+	H⁺	→	2,2′-bipyridineH⁺
Initial mmol:	1.41₉		x		—
Final mmol:	1.41₉ − x		—		x

$pH = pK_a + \log \frac{[\text{bipyridine}]}{[\text{bipyridineH}^+]}$

$4.19 = 4.34 + \log \frac{1.41_9 - x}{x} \Rightarrow x = 0.831$ mmol

volume = $\frac{0.831 \text{ mmol}}{0.246 \text{ mmol/mL}} = 3.38$ mL

9-14. 5.00 g of HEPES = 0.0210 mol HA. We will add x mol OH⁻ to get a pH of 7.40.

	HA	+	OH⁻	→	A⁻	+	H₂O
Initial moles:	0.021 0		x		—		
Final moles:	0.021 0 − x		—		x		

$$\text{pH} = 7.40 = \text{p}K_a + \log \frac{[A^-]}{[HA]} = 7.56 + \log \frac{x}{0.021\,0 - x} \Rightarrow x = 8.59 \times 10^{-3} \text{ mol}$$

$$\text{volume} = \frac{8.59 \times 10^{-3} \text{ mol}}{0.626 \text{ M}} = 13.7 \text{ mL}$$

9-15. 52.2 mL of 0.013 4 M morpholine = 0.699_5 mmol base. We will add x mol H^+ to the morpholine to get a pH of 8.00.

	B	+	H^+	→	BH^+
Initial mmol:	0.699_5		x		—
Final mmol:	$0.699_5 - x$		—		x

The Henderson-Hasselbalch equation for this solution is

$$8.00 = 8.492 + \log \frac{0.699_5 - x}{x} \Rightarrow x = 0.529\,1 \text{ mmol } H^+$$

mL required = (0.529 1 mmol)/(0.113 mmol/mL) = 4.68 mL H^+

9-16. Buffer capacity is maximum when pH = $\text{p}K_a$ for the buffer. We look for buffers for which $\text{p}K_a$ is closest to the desired pH.

(a) For pH 4.00, use citric acid or acetic acid.

(b) For pH 7.00, use imidazole hydrochloride.

(c) For pH 10.00, use CAPS.

(d) Other buffers whose $\text{p}K_a$ is within ±1 unit of pH are also useful. For pH 10.00, CHES, boric acid, or ammonia would work, but they would not have as great a buffer capacity as CAPS.

9-17. The $\text{p}K_a$ values are (i) 10.77, (ii) 9.24, (iii) 5.96, and (iv) 7.15. $\text{p}K_a$ for ammonia is closest to 9.0, so buffer (ii) has the greatest buffer capacity at pH 9.0.

9-18. (a) HEPES is an acid, so we need NaOH to bring the pH up to 7.45.

(b) 1. Weigh out (0.250 L)(0.050 0 M) = 0.012 5 mol of HEPES and dissolve in ~200 mL.

  2. Adjust pH to 7.45 with NaOH (HEPES is an acid).

  3. Dilute to 250 mL.

9-19. (a) 1. Weigh out (0.500 L)(0.100 M) = 0.012 5 mol of imidazole hydrochloride and dissolve in ~400 mL.

  2. Adjust pH to 7.50 with NaOH (because imidazole hydrochloride is an acid).

  3. Dilute to 500 mL.

Buffers

(b) If we started with imidazole instead of imidazole hydrochloride, we would need HCl to adjust the pH because imidazole is a base.

9-20. (a) 250.0 mL of 0.100 M buffer requires 0.025 0 mol NaOAc · 2H$_2$O = 2.95 g. We will add x mol strong acid to convert x mol NaOAc into x mol HOAc.

	OAc$^-$	+	H$^+$	→	HOAc
Initial mol:	0.025 0		x		—
Final mol:	0.025 0 – x		10$^{-5.00}$		x

$$pH = pK_a + \log \frac{[OAc^-]}{[HOAc]}$$

$$5.00 = 4.756 + \log \frac{0.025\,0 - x}{x} \Rightarrow x = 9.08 \text{ mmol} = 90.8 \text{ mL HCl}$$

(b) 1. Dissolve 2.95 g NaOAc · 2H$_2$O in ~100 mL H$_2$O.
2. While measuring the pH, add enough HCl to bring the pH to 5.00.
3. Dilute to exactly 250 mL.

9-21. (a) red (b) orange (c) yellow (d) red

9-22. (a) $pH = pK_a + \log \frac{[A^-]}{[HA]}$

$$6.200 = 5.000 + \log \frac{1-q}{q}$$

$$1.200 = \log \frac{1-q}{q} \Rightarrow 10^{1.200} = 10^{\log[(1-q)/q]}$$

$$15.85 = \frac{1-q}{q} \Rightarrow q = 0.059\,35 \text{ mol and } p = 1 - q = 0.940\,6 \text{ mol}$$

(b)

	A$^-$	+	H$^+$	→	HA
Initial mol:	0.940 6		0.010 0		0.059 35
Final mol:	0.930 6		—		0.069 35

$$pH = 5.000 + \log \frac{0.930\,6}{0.069\,35} = 6.128$$

$\Delta(pH) = 6.128 - 6.200 = -0.072$ (which agrees with the figure)

9-23. The pH will shift by the amount that pK_a shifts. For ammonia, Table 9-2 shows $\Delta(pK_a)/\Delta T = -0.031$ per degree near 25°C.

$$pK_a \text{ (at 37°C)} = pK_a \text{ (at 25°C)} + [\Delta(pK_a)/\Delta T](\Delta T)$$
$$= 9.24 + (-0.031°C^{-1})(12°C) = 8.87$$

Because pK_a changes by –0.37, we expect pH to change by –0.37.

Because the pH at 25° was 9.50, the pH at 37° will be 9.50 – 0.37 = 9.13.

9-24. Henderson-Hasselbalch spreadsheet

	A	B	C	D
1	pKa =	[A-]/[HA]	pH	log([A-]/[HA])
2	4	0.001	1.00	−3.0000
3		0.01	2.00	−2.0000
4		0.05	2.70	−1.3010
5		0.5	3.70	−0.3010
6		1	4.00	0.0000
7		2	4.30	0.3010
8		5	4.70	0.6990
9		10	5.00	1.0000
10		100	6.00	2.0000
11		1000	7.00	3.0000
12				
13	C2 = $A$2 + LOG(B2)			
14	D2 = LOG(B2)			

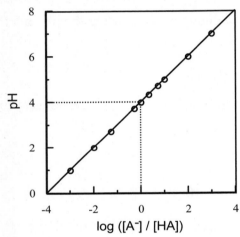

The Henderson-Hasselbalch equation says that pH = pK_a + log([A$^-$]/[HA]). A graph of pH vs log([A$^-$]/[HA]) should be a straight line with a slope of 1. When log([A$^-$]/[HA]) = 0, pH should be equal to pK_a.

9-25. Let the indicator equilibrium be abbreviated HIn$^-$ $\overset{K_2}{\rightleftharpoons}$ In^{2-} + H$^+$, for which the Henderson-Hasselbalch equation is

$$pH = pK_2 + \log \frac{[\text{In}^{2-}]}{[\text{HIn}^-]}$$

If pK_2 is known accurately, we can find the pH of the solution by accurately measuring the quotient [In^{2-}]/[HIn$^-$]. This quotient is available from measuring the visible absorbance of the solution at the two wavelengths of maximum absorbance. Regardless of what other equilibria are present in the solution, knowing the quotient [In^{2-}]/[HIn$^-$] establishes the pH.

CHAPTER 10
ACID-BASE TITRATIONS

10-1. Prior to the equivalence point, the reaction is $OH^- + H^+ \rightarrow H_2O$, and the pH is computed from the concentration of excess OH^-. At the equivalence point, the solution contains just neutral salt in water, and the pH is 7. After the equivalence point, there is excess H^+ present, and the pH is determined by the concentration of excess H^+.

10-2. The initial solution contains a known concentration of the weak acid, and the pH is determined from the acid dissociation reaction: $HA \rightarrow H^+ + A^-$. Between the initial point and the equivalence point, each increment of OH^- neutralizes a corresponding increment of HA, leaving a known mixture of HA and A^-. *Aha! A buffer!* The pH is given by $pH = pK_a + \log([A^-]/[HA])$. At the equivalence point, the solution contains the weak base, A^-, at a concentration more dilute than that of the original HA. The pH is determined by the base hydrolysis reaction: $A^- + H_2O \rightarrow HA + OH^-$. Beyond the equivalence point, there is excess OH^- titrant in the solution. The pH is determined from the concentration of excess OH^-.

10-3. The initial solution contains a known concentration of the weak base, and the pH is determined from the hydrolysis reaction: $A^- + H_2O \rightarrow HA + OH^-$. Between the initial point and the equivalence point, each increment of H^+ neutralizes a corresponding increment of A^-, thereby leaving a known mixture of HA and A^-. *Aha! A buffer!* The pH is given by $pH = pK_a + \log([A^-]/[HA])$. At the equivalence point the solution contains the weak acid, HA, at a concentration more dilute than that of the original A^-. The pH is determined by the acid dissociation reaction: $HA \rightarrow H^+ + A^-$. Beyond the equivalence point, there is excess H^+ titrant in the solution. The pH is determined from the concentration of excess H^+.

10-4. Immediately prior to the equivalence point, the pH is controlled by the buffer action of the weak base, B, and its conjugate acid, BH^+. As we get very close to the equivalence point, almost all the base is gone and the concentration of BH^+ is essentially constant. The log term in the Henderson-Hasselbalch equation, $[B]/[BH^+]$, changes very rapidly with each tiny increment of added H^+. At the equivalence point, the concentration of H^+ is $\sim 10^{-6.5}$ M. Each subsequent addition of strong acid from the buret increases $[H^+]$ by a relatively large amount, and the pH continues to drop rapidly until there is a substantial concentration of excess strong acid.

10-5. The equivalence point is the steepest point in the titration curve. The maximum in the first derivative occurs at the steepest point of the original titration curve. When the first derivative reaches a maximum, the second derivative (which is the derivative of the first derivative) is zero.

10-6. Find V_e: $(V_e)(1.00\ M) = (100.0\ mL)(0.100\ M) \Rightarrow V_e = 10.0\ mL$

V_a:	0	1	5	9	9.9	10	10.1	12
pH:	13.00	12.95	12.68	11.96	10.96	7.00	3.04	1.75

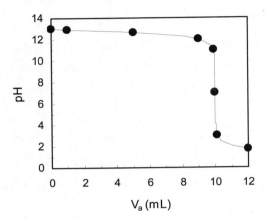

Representative calculations:

0 mL: $\quad pH = -\log\dfrac{K_w}{[OH^-]} = -\log\dfrac{10^{-14}}{0.100} = 13.00$

1.00 mL: $\quad mol\ OH^- = \underbrace{(100.0\ mL)(0.100\ M)}_{\text{Initial mmol OH}^-} - \underbrace{(1.00\ mL)(1.00\ M)}_{\text{Added mmol H}^+} = 9.00\ mmol$

$\qquad\qquad [OH^-] = (9.00\ mmol)/(101\ mL) = 0.0891\ M \Rightarrow pH = 12.95$

10.00 mL: $\quad [OH^-] = [H^+] = 1.0 \times 10^{-7}\ M \Rightarrow pH = 7.00$

10.10 mL: $\quad [H^+] = \dfrac{(0.10\ mL)(1.00\ M)}{(110.1\ mL)} = 9.08 \times 10^{-4}\ M \Rightarrow pH = 3.04$

10-7. Find V_e: $(V_e)(0.100\ M) = (25.0\ mL)(0.0500\ M) \Rightarrow V_e = 12.5\ mL$

0 mL: $\quad pH = -\log(0.0500) = 1.30$

1.00 mL: $\quad mol\ H^+ = \underbrace{(25.0\ mL)(0.0500\ M)}_{\text{Initial mmol H}^+} - \underbrace{(1.00\ mL)(0.100\ M)}_{\text{Added mmol OH}^-} = 1.15\ mmol$

$\qquad\qquad [H^+] = (1.15\ mmol)/(26.0\ mL) = 0.0442\ M \Rightarrow pH = 1.35$

5.00 mL: $\quad pH = 1.60$

10.00 mL: $\quad pH = 2.15$

12.40 mL: $\quad pH = 3.57$

Acid-Base Titrations

12.50 mL: equivalence point ⇒ pH = 7.00

12.60 mL: excess $[OH^-] = \dfrac{(0.1\text{ mL})(0.100\text{ M})}{37.6\text{ mL}} = 2.66 \times 10^{-4}\text{ M} \Rightarrow$ pH = 10.42

13.00 mL: pH = 11.12

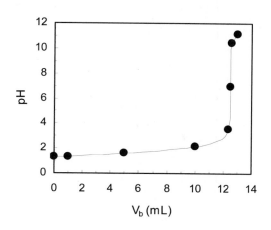

10-8. Titration reaction: $HA + OH^- \rightarrow A^- + H_2O$

Find V_e: $(V_e)(0.500\text{ M}) = (50.0\text{ mL})(0.0500\text{ M}) \Rightarrow V_e = 5.00\text{ mL}$

0 mL: $HA \rightleftharpoons H^+ + A^-$

$\dfrac{x^2}{0.0500 - x} = 10^{-4.00} \Rightarrow x = 2.19 \times 10^{-3}\text{ M} \Rightarrow$ pH = 2.66

1.00 mL:

	HA	+	OH⁻	→	A⁻	+	H₂O
Initial mmol:	2.50		0.50		—		
Final mmol:	2.00		—		0.50		

$\text{pH} = \text{p}K_a + \log\dfrac{[A^-]}{[HA]} = 4.00 + \log\dfrac{0.50}{2.00} = 3.40$

2.50 mL: pH $= 4.00 + \log\left(\dfrac{1.25}{1.25}\right) = 4.00$

4.00 mL: pH $= 4.00 + \log\left(\dfrac{2.00}{0.50}\right) = 4.60$

4.90 mL: pH $= 4.00 + \log\left(\dfrac{2.45}{0.05}\right) = 5.69$

5.00 mL: $A^- + H_2O \rightleftharpoons HA + OH^-$ $\text{p}K_b = 10.00$
$\phantom{5.00\text{ mL: }}0.0455 - x x x$

$F' = \dfrac{(50.0\text{ mL})(0.0500\text{ M})}{55.0\text{ mL}} = 0.0455\text{ M}$

$\dfrac{x^2}{0.0455 - x} = 1.00 \times 10^{-10} \Rightarrow x = [OH^-] = 2.13 \times 10^{-6} \Rightarrow$ pH = 8.33

5.10 mL: $[OH^-] = \dfrac{(0.10\ \text{mL})(0.500\ M)}{55.10\ \text{mL}} = 9.07 \times 10^{-4}\ M \Rightarrow \text{pH} = 10.96$

6.00 mL: $[OH^-] = \dfrac{(1.00\ \text{mL})(0.500\ M)}{56.0\ \text{mL}} = 8.93 \times 10^{-3}\ M \Rightarrow \text{pH} = 11.95$

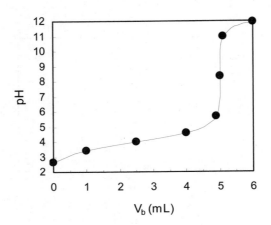

10-9. $\underbrace{(V_e\ (\text{mL}))(0.0500\ M)}_{\text{mmol OH}^-} = \underbrace{(25.0\ \text{mL})(0.0100\ M)}_{\text{mmol A}} \Rightarrow V_e = 5.00\ \text{mL}$

$V_b = 0$ mL: HA $\rightleftharpoons$ H$^+$ + A$^-$ $K_a = 2.33 \times 10^{-11}$, p$K_a = 10.645$
     $0.0100-x$ x x

$\dfrac{x^2}{0.0100-x} = K_a \Rightarrow x = 4.83 \times 10^{-7} \Rightarrow \text{pH} = -\log x = 6.32$

$V_b = 2.50$ mL: HA + OH$^-$ → A$^-$ + H$_2$O

Initial mmol: 0.250   0.125   —
Final mmol: 0.125   —   0.125

$\text{pH} = \text{p}K_a + \log \dfrac{[A^-]}{[HA]} = 10.632 + \log \dfrac{0.125}{0.125} = 10.63$

$V_b = V_e = 5.00$ mL: Formal concentration of A$^-$ = $F' = \dfrac{\text{mmol A}^-}{\text{total mL}}$

$= \dfrac{(25.0\ \text{mL})(0.0100\ M)}{30.0\ \text{mL}} = 8.33 \times 10^{-3}\ M$

A$^-$ + H$_2$O $\rightleftharpoons$ HA + OH$^-$ $K_b = K_w/K_a = 4.29 \times 10^{-4}$
$F'-x$ x x

$\dfrac{x^2}{F'-x} = K_b \Rightarrow x = 1.69 \times 10^{-3} \Rightarrow \text{pH} = -\log \dfrac{K_w}{x} = 11.23$

Acid-Base Titrations

$V_b = 10.00$ mL: excess $[OH^-] = \dfrac{(5.0 \text{ mL})(0.0500 \text{ M})}{35.0 \text{ mL}} = 7.14 \times 10^{-3}$ M

$\Rightarrow$ pH = 11.85

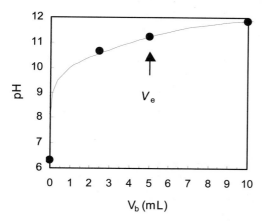

The smooth curve for the graph was calculated exactly by using a spreadsheet, as described later in this chapter. There is no break at the equivalence point because pK_a for the acid BH^+ is too high—which is another way of saying that the acid is just too weak.

10-10. Titration reaction: C₆H₅–NH_3^+ + OH^- → C₆H₅–NH_2 + H_2O; $V_e = 100$ mL

BH^+ ($pK_a = 4.601$)   B

0 mL: $BH^+ \rightleftharpoons B + H^+$ $K_a = 2.51 \times 10^{-5}$
       $0.100 - x$ x x

$\dfrac{x^2}{0.100 - x} = K_a \Rightarrow x = 1.57 \times 10^{-3}$ M $\Rightarrow$ pH = 2.80

$0.100\, V_e = 10.0$ mL: $BH^+ + OH^- \rightarrow B + H_2O$

Initial mmol:	10.00	1.00	—
Final mmol:	9.00	—	1.00

pH = pK_a (for BH^+) + $\log \dfrac{[B]}{[BH^+]} = 4.601 + \log \dfrac{1.00}{9.00} = 3.65$

$0.500\, V_e = 50.0$ mL: pH = pK_a = 4.60

$0.900\, V_e = 90.0$ mL: pH = $4.601 + \log \dfrac{9.00}{1.00} = 5.56$

$V_e = 100.0$ mL: F' = [B] = $\dfrac{10.0 \text{ mmol}}{200 \text{ mL}} = 0.0500$ M

     B + H_2O ⇌ BH^+ + OH^- $K_b = \dfrac{K_w}{K_a} = 3.98 \times 10^{-10}$
     F' − x   x   x

$\dfrac{x^2}{F' - x} = K_b \Rightarrow x = 4.46 \times 10^{-6}$ M

$[H^+] = \dfrac{K_w}{x} = 2.24 \times 10^{-9} \Rightarrow$ pH = 8.65

1.200 V_e = 120.0 mL: excess [OH$^-$] = $\dfrac{(20.0\ \text{mL})(0.100\ \text{M})}{220\ \text{mL}}$ = 9.09 × 10^{-3} M

$\Rightarrow$ pH = 11.96

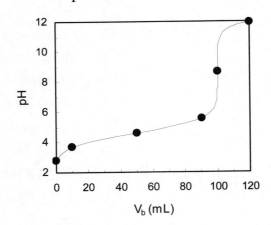

10-11. The titration reaction is HA + OH$^-$ → A$^-$ + H$_2$O. A volume of V mL of HA will require $2V$ mL of KOH to reach the equivalence point, because

$(V)(0.100\ \text{M}) = (V_e)(0.0500\ \text{M}) \Rightarrow V_e = 2.00V$

Formal concentration of A$^-$ at equivalence point = $\dfrac{(V)(0.100\ \text{M})}{V+2V}$ = 0.033 3 M

The pH is found by writing

A$^-$ $\rightleftharpoons$ H$_2$O = HA + OH$^-$ $\dfrac{x^2}{0.033\,3-x} = K_b = \dfrac{K_w}{K_a} = 6.76 \times 10^{-11}$
0.033 3 $-x$ x x

$\Rightarrow x = 1.50 \times 10^{-6}$ M $\Rightarrow$ pH = 8.18

10-12. Find V_e: $(V_e\ (\text{mL}))(0.0643\ \text{M}) = (25.00\ \text{mL})(0.0938\ \text{M}) \Rightarrow V_e$ = 36.5 mL

At V_b = 16.24 mL, we are in the buffer region before the equivalence point.

	HA	+	OH$^-$	→	A$^-$	+	H$_2$O
Initial mmol:	2.345		1.044		—		
Final mmol:	1.301		—		1.044		

$3.62 = pK_a + \log \dfrac{1.044}{1.301} \Rightarrow pK_a = 3.72$

Acid-Base Titrations

10-13.

	HA	+	OH$^-$	$\rightarrow$	A$^-$	+	H$_2$O
Initial mmol:	5.857		x		—		
Final mmol:	5.857$-x$		—		x		

$$\text{pH} = 9.13 = pK_a + \log \frac{[\text{A}^-]}{[\text{HA}]} = 9.39 + \log \frac{x}{5.857-x} \Rightarrow x = 2.077 \text{ mmol}$$

$$[\text{OH}^-] = \frac{2.077 \text{ mmol}}{22.63 \text{ mL}} = 0.0918 \text{ M}$$

10-14. (a) Titration reaction: HA + OH$^-$ $\rightarrow$ A$^-$ + H$_2$O

mmol HA = mmol OH$^-$ required to reach V_e = (27.63 mL)(0.09381 M)
= 2.592 mmol. The concentration of HA is (2.592 mmol/100.0 mL)
= 0.02592 M.

(b) At the equivalence point, the initial volume of 100.0 mL of 0.02592 M HA has been diluted up to 127.63 mL and converted to A$^-$. The formal concentration of A$^-$ is $(0.02592 \text{ M})\left(\frac{100.0 \text{ mL}}{127.63 \text{ mL}}\right) = 0.02031$ M

(c) Because the pH is 10.99, [OH$^-$] = 9.77 × 10^{-4}, and we can write

	A$^-$	+	H$_2$O	$\rightleftharpoons$	HA	+	OH$^-$
	(0.02031 $-$ 9.77 × 10^{-4})				9.77 × 10^{-4}		9.77 × 10^{-4}

$$K_b = \frac{[\text{HA}][\text{OH}^-]}{[\text{A}^-]} = \frac{(9.77 \times 10^{-4})^2}{0.02031 - (9.77 \times 10^{-4})} = 4.94 \times 10^{-5}$$

$$K_a = \frac{K_w}{K_b} = 2.03 \times 10^{-10} \Rightarrow pK_a = 9.69$$

(d) For the 19.47-mL point, we have

	HA	+	OH$^-$	$\rightarrow$	A$^-$	+	H$_2$O
Initial mmol:	2.592		1.826		—		
Final mmol:	0.766		—		1.826		

$$\text{pH} = pK_a + \log \frac{[\text{A}^-]}{[\text{HA}]} = 9.69 + \log \frac{1.826}{0.766} = 10.07$$

10-15. Titration reaction: B + H$^+$ $\rightarrow$ BH$^+$ $K_b = 10^{-5.00}$ V_e = 10.0 mL

V_a:	0	1.00	5.00	9.00	9.90	10.00	10.10	12.00
pH:	11.00	9.95	9.00	8.05	7.00	5.02	3.04	1.75

Representative calculations:

0 mL: $B + H_2O \rightleftharpoons BH^+ + OH^-$

$ 0.100-x x x$

$\dfrac{x^2}{0.100-x} = 10^{-5.00}$ M $\Rightarrow x = 9.95 \times 10^{-4}$

$[H^+] = \dfrac{K_w}{x} \Rightarrow$ pH = 11.00

1.00 mL:

	B	+	H^+	→	BH^+
Initial mmol:	10.0		1.00		—
Final mmol:	9.0		—		1.00

$\text{pH} = \text{p}K_{BH^+} + \log \dfrac{[B]}{[BH^+]} = 9.00 + \log \dfrac{9.0}{1.0} = 9.95$

10.00 mL: $BH^+ \rightleftharpoons B + H^+$

$\dfrac{(100 \text{ mL})(0.100-x)}{110 \text{ mL}} x x$

$\dfrac{x^2}{0.0909-x} = \dfrac{K_w}{K_b} \Rightarrow x = 9.53 \times 10^{-6}$

$\Rightarrow$ pH = 5.02

10.10 mL: excess $[H^+] = \dfrac{(0.1 \text{ mL})(1.00 \text{ M})}{110.1 \text{ mL}} = 9.08 \times 10^{-4}$ M $\Rightarrow$ pH = 3.04

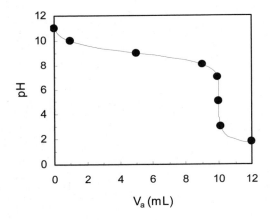

10-16. Titration reaction: $CH_3CH_2CO_2^- + H^+ \rightarrow CH_3CH_2CO_2H$

Find V_e: $\underbrace{(V_e)(0.0837 \text{ M})}_{\text{mmol HCl}} = \underbrace{(100.0 \text{ mL})(0.0400 \text{ M})}_{\text{mmol NaA}} \Rightarrow V_e = 47.79$ mL

0 mL: $A^- + H_2O \rightleftharpoons HA + OH^-$ $K_b = \dfrac{K_w}{K_a} = 7.46 \times 10^{-10}$

$ 0.0400-x x x$

$\dfrac{x^2}{0.0400-x} = K_b \Rightarrow x = 5.47 \times 10^{-6}$ M pH $= -\log \dfrac{K_w}{x} = 8.74$

Acid-Base Titrations

$1/4\ V_e$:

	A^-	$+$	H^+	$\rightarrow$	HA
Initial mmol:	4.00		1.00		—
Final mmol:	3.00		—		1.00

$$pH = pK_a + \log\frac{[A^-]}{[HA]} = 4.874 + \log\frac{3.00}{1.00} = 5.35$$

$1/2\ V_e$: $\quad pH = pK_a = 4.87$

$3/4\ V_e$: $\quad pH = pK_a + \log\dfrac{1.00}{3.00} = 4.40$

V_e: $\quad HA \rightleftharpoons H^+ + A^-$
$\quad\quad\quad F-x \quad\ x \quad\ x$

$$F_{HA} = \frac{(100.0\ \text{mL})(0.040\,0\ M)}{147.79\ \text{mL}} = 0.027\,1\ M$$

$$\frac{x^2}{0.027\,1 - x} = K_a = 1.34 \times 10^{-5} \Rightarrow x = 5.96 \times 10^{-4}\ M \Rightarrow pH = 3.22$$

$1.1\ V_e = 52.57\ \text{mL}$: $[H^+] = \dfrac{(4.779\ \text{mL})(0.083\,7\ M)}{152.6\ \text{mL}} = 2.62 \times 10^{-3}\ M \Rightarrow pH = 2.58$

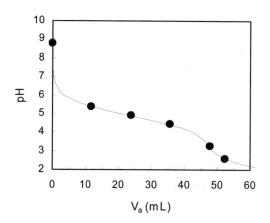

10-17. (a) C$_6$H$_5$—CH$_2$NH$_2$ + H$^+$ $\rightleftharpoons$ C$_6$H$_5$—CH$_2$NH$_3^+$ $\quad pK_a = 9.35$

$K = 1/K_a$ (for $C_6H_5CH_2NH_3^+$) $= 2.2 \times 10^9$

(b) Titration reaction: $B + H^+ \rightarrow BH^+$. To find the equivalence point, we write
$(50.0\ \text{mL})(0.031\,9\ M) = (V_e)(0.050\,0\ M) \Rightarrow V_e = 31.9\ \text{mL}$

0 mL: $\quad B + H_2O \rightleftharpoons BH^+ + OH^-$
$\quad\quad\quad\ \ 0.031\,9 - x \quad\quad\ \ x \quad\ \ x$

$$\frac{x^2}{0.031\,9 - x} = K_b = \frac{K_w}{K_a} \Rightarrow 2.22 \times 10^{-5}$$

$$\Rightarrow x = 8.31 \times 10^{-4}\ M \Rightarrow pH = 10.92$$

12.0 mL:

	B	+	H⁺	→	BH⁺
Initial mmol:	1.595		0.600		—
Final mmol:	0.995		—		0.600

$$\text{pH} = \text{p}K_a + \log\frac{[\text{B}]}{[\text{BH}^+]} = 9.35 + \log\frac{0.995}{0.600} = 9.57$$

$\frac{1}{2}V_e$: $\text{pH} = \text{p}K_a = 9.35$

30.0 mL:

	B	+	H⁺	→	BH⁺
Initial mmol:	1.595		1.500		—
Final mmol:	0.095		—		1.500

$$\text{pH} = \text{p}K_a + \log\frac{[\text{B}]}{[\text{BH}^+]} = 9.35 + \log\frac{0.095}{1.500} = 8.15$$

V_e: B has been converted to BH⁺ at a concentration of

$$\frac{(50.0\text{ mL})(0.031\,9\text{ M})}{81.9\text{ mL}} = 0.019\,5\text{ M}$$

$$\text{BH}^+ \rightleftharpoons \text{B} + \text{H}^+$$
$$0.0195 - x \quad x \quad x$$

$$\frac{x^2}{0.019\,5 - x} = K_a = 4.5 \times 10^{-10}$$

$$\Rightarrow x = 2.96 \times 10^{-6}\text{ M} \Rightarrow \text{pH} = 5.53$$

35.0 mL: $\quad\text{excess }[\text{H}^+] = \dfrac{(3.1\text{ mL})(0.050\,0\text{ M})}{85.0\text{ mL}} = 1.82 \times 10^{-3}\text{ M}$

$\Rightarrow \text{pH} = 2.74$

10-18. Titration reaction: $\text{CN}^- + \text{H}^+ \rightarrow \text{HCN}$

At the equivalence point, moles of CN^- = moles of H^+.

$(0.100\text{ M})(50.00\text{ mL}) = (0.438\text{ M})(V_e) \Rightarrow V_e = 11.42\text{ mL}$

(a)

	CN⁻	+	H⁺	→	HCN
Initial mmol:	5.00		1.84		—
Final mmol:	3.16		—		1.84

$$\text{pH} = \text{p}K_a + \log\frac{[\text{CN}^-]}{[\text{HCN}]} = 9.21 + \log\frac{3.16}{1.84} = 9.44$$

(b) 11.82 mL is 0.40 mL past the equivalence point.

$$\text{Excess }[\text{H}^+] = \frac{(0.40\text{ mL})(0.438\text{ M})}{61.82\text{ mL}} = 2.83 \times 10^{-3}\text{ M} \Rightarrow \text{pH} = 2.55$$

(c) At the equivalence point we have made HCN at a formal concentration of

$$\frac{(50.00\text{ mL})(0.100\text{ M})}{61.42\text{ mL}} = 0.081\,4\text{ M}$$

Acid-Base Titrations

$$\text{HCN} \rightleftharpoons \text{H}^+ + \text{CN}^- \qquad \frac{x^2}{0.0814-x} = K_a = 6.2 \times 10^{-10}$$
$$\quad 0.0814-x \qquad x \qquad x$$
$$\Rightarrow x = 7.10 \times 10^{-6} \Rightarrow \text{pH} = 5.15$$

10-19. Titration reaction: B (Imidazole) + H$^+$ → BH$^+$

$$K_{BH^+} = 1.02 \times 10^{-7}$$
$$pK_{BH^+} = 6.993$$

At the equivalence point, moles of imidazole = moles of H$^+$.
(0.050 00 M)(25.00 mL) = (0.125 0 M)(V_e) ⇒ V_e = 10.00 mL

V_a:	0	1.00	5.00	9.00	9.90	10.00	10.10	12.00
pH:	9.85	7.95	6.99	6.04	5.00	4.22	3.45	2.17

Representative calculations:

0 mL: $\quad$ B + H$_2$O ⇌ BH + OH$^-$ $\qquad \frac{x^2}{0.050\,00-x} = K_b = K_w/K_a = 9.8 \times 10^{-8}$
$\qquad\quad 0.050\,00-x \qquad x \quad x$
$$\Rightarrow x = 7.00 \times 10^{-5} \text{ M}$$
$$[\text{H}^+] = \frac{K_w}{x} \Rightarrow \text{pH} = 9.85$$

1.00 mL: $\qquad$ B + H$^+$ → BH$^+$

	B	H$^+$	BH$^+$
Initial mmol:	1.250	0.125	—
Final mmol:	1.125	—	0.125

$$\text{pH} = pK_{BH^+} + \log\frac{[\text{B}]}{[\text{BH}^+]} = 6.993 + \log\frac{1.125}{0.125} = 7.95$$

10.00 mL: $\qquad$ BH$^+$ ⇌ B + H$^+$ $\qquad \frac{x^2}{0.035\,7-x} = K_{BH^+} = 1.02 \times 10^{-7}$
$\qquad\quad \frac{(25\text{ mL})(0.05\text{ M})}{35\text{ mL}} - x \quad x \quad x$
$$\Rightarrow x = 6.03 \times 10^{-5} \Rightarrow \text{pH} = 4.22$$

12.00 mL: $\quad$ excess [H$^+$] = $\frac{(2.00 \text{ mL})(0.125 \text{ M})}{37.00 \text{ mL}}$ = 6.76 × 10^{-3} M ⇒ pH = 2.17

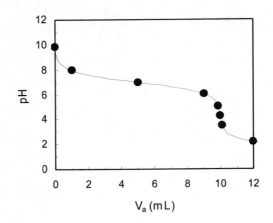

10-20. No. When a weak acid is titrated with a strong base, the solution contains A^- at the equivalence point. A solution of A^- must have a pH above 7.

10-21. Yellow, green, blue

10-22. There is not an abrupt change in pH at V_e. The color of the indicator would change very gradually.

10-23. (a) Colorless → pink

(b) If you do the experiment too slowly, the pink keeps fading and you keep adding more NaOH to turn the solution pink again. There is a systematic error in which it appears that more NaOH is required than should be.

10-24. Two candidates: cresol red (orange → red) and phenolphthalein (colorless → pink)

10-25.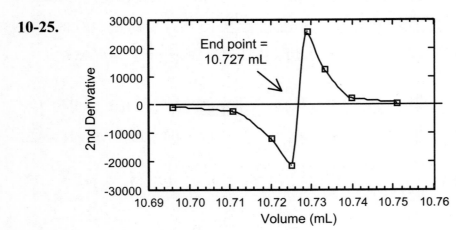

Acid-Base Titrations

	A	B	C	D	E	F
1	Derivatives in a titration curve					
2	Data		1st derivative		2nd derivative	
3	mL NaOH	pH	mL	ΔpH/ΔmL	mL	Δ(ΔpH/ΔmL)
4	10.679	7.643				ΔmL
5			10.6875	-11.5		
6	10.696	7.447			10.6960	-553.6
7			10.7045	-20.9		
8	10.713	7.091			10.7108	-2234.7
9			10.7170	-48.9		
10	10.721	6.700			10.7200	-11770.8
11			10.7230	-119.5		
12	10.725	6.222			10.7250	-21375.0
13			10.7270	-205.0		
14	10.729	5.402			10.7290	25687.5
15			10.7310	-102.2		
16	10.733	4.993			10.7333	12411.1
17			10.7355	-46.4		
18	10.738	4.761			10.7398	2351.0
19			10.7440	-26.4		
20	10.750	4.444			10.7508	885.2
21			10.7575	-14.5		
22	10.765	4.227				
23	Representative formulas:					
24	C5 = (A6+A4)/2			E6 = (C7+C5)/2		
25	D5 = (B6-B4)/(A6-A4)			F6 = (D7-D5)/(C7-C5)		

10-26. One mole of borax reacts with two moles of H^+:

$$[HNO_3](0.02161\ L) = 2\left(\frac{0.2619\ g\ borax}{381.37\ g\ borax/mol}\right) \Rightarrow [HNO_3] = 0.06356\ M$$

$\underbrace{}_{mol\ H^+}$ $\underbrace{\phantom{\frac{0.2619\ g\ borax}{381.37\ g\ borax/mol}}}_{mol\ borax}$

10-27. (a) Fraction of cleaner analyzed $= \dfrac{4.373\ g}{10.231\ g + 39.466\ g} = 0.08799_3$

(b) mol NH_3 present = mol HCl used = $(0.01422\ L)(0.1063\ M) = 1.511_6$ mmol

$(1.511_6 \times 10^{-3}\ mol\ NH_3)(17.031\ g/mol) = 25.74_4\ mg\ NH_3$

(c) grams of cleaner titrated = $\underbrace{(0.08799_3)}_{\substack{Fraction\ of\\sample\ titrated}} \underbrace{(10.231\ g)}_{\substack{Initial\\sample}} = 0.900\ 26\ g$

$$wt\%\ NH_3 = \frac{2.574_4 \times 10^{-2}\ g\ NH_3}{0.900\ 2_6\ g\ sample} \times 100 = 2.860\ wt\%$$

10-28. (a) $NH_4^+ \rightleftharpoons NH_3 + H^+$ $\quad\quad \dfrac{x^2}{0.010-x} = K_a = 5.69 \times 10^{-10}$

$0.010-x \quad\quad x \quad\quad x$

$\Rightarrow x = 2.38 \times 10^{-6}\ M \Rightarrow pH = 5.62$

(b) One possible indicator is methyl red, using the yellow end point.

10-29. Titration of weak base with strong acid

	A	B	C	D	E	F	G
1	Ca =	pH	[H+]	[OH-]	Alpha(BH+)	Phi	Va (mL)
2	0.05	10.49	3.24E-11	3.09E-04	3.135E-02	4.394E-04	0.009
3	Cb =	10.00	1.00E-10	1.00E-04	9.091E-02	8.075E-02	1.615
4	0.01	9.00	1.00E-09	1.00E-05	5.000E-01	4.989E-01	9.978
5	Vb =	8.00	1.00E-08	1.00E-06	9.091E-01	9.090E-01	18.179
6	100	7.00	1.00E-07	1.00E-07	9.901E-01	9.901E-01	19.802
7	Kb =	6.00	1.00E-06	1.00E-08	9.990E-01	9.991E-01	19.982
8	0.00001	5.00	1.00E-05	1.00E-09	9.999E-01	1.001E+00	20.022
9	K(BH+) =	4.00	1.00E-04	1.00E-10	1.000E+00	1.012E+00	20.240
10	1.E-09	3.00	1.00E-03	1.00E-11	1.000E+00	1.122E+00	22.449
11	Kw =	2.75	1.78E-03	5.62E-12	1.000E+00	1.221E+00	24.425
12	1.E-14						
13							
14	A10 = 1E-14/$A$8				C2 = 10^-B2		
15	D2 = $A$12/C2				E2 = C2/(C2 + $A$10)		
16	F2 = (E2 + (C2 − D2)/$A$4)/(1 − (C2 − D2)/$A$2) [Equation 10-15]						
17	G2 = F2*$A$4*$A$6/$A$2						

10-30. Titration of weak base with strong acid

	A	B	C	D	E	F	G
1	Ca =	pH	[H+]	[OH-]	Alpha(BH+)	Phi	Va (mL)
2	0.1	2.00	1.00E-02	1.00E-12	9.90E-01	1.66E+00	16.557
3	Cb =	2.90	1.26E-03	7.94E-12	9.26E-01	1.00E+00	10.020
4	0.02	3.50	3.16E-04	3.16E-11	7.60E-01	7.78E-01	7.780
5	Vb =	4.00	1.00E-04	1.00E-10	5.00E-01	5.06E-01	5.055
6	50	4.50	3.16E-05	3.16E-10	2.40E-01	2.42E-01	2.419
7	K(BH+) =	6.00	1.00E-06	1.00E-08	9.90E-03	9.95E-03	0.100
8	1E-04	8.15	7.08E-09	1.41E-06	7.08E-05	5.17E-07	0.000
9	Kw =						
10	1E-14			E2 = C2/(C2 + $A$8)			
11		C2 = 10^-B2		F2 = (E2 + (C2 − D2)/$A$4)/(1 − (C2 −D2)/$A$2)			
12		D2 = $A$10/C2		G2 = F2*$A$4*$A$6/$A$2			

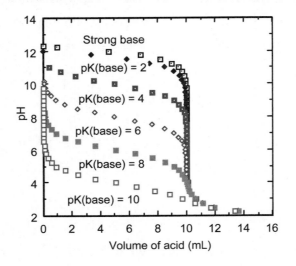

Acid-Base Titrations

10-31. In the coarse titration curve below, there appear to be two end points at about 27.5 and 41.5 mL.

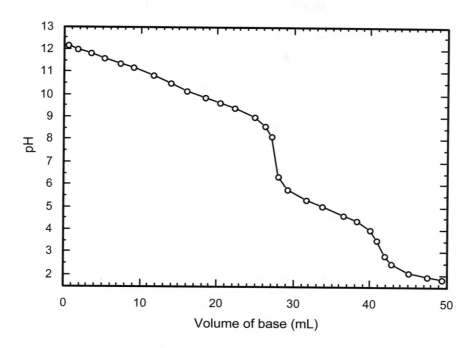

The first and second derivatives near the first end point are calculated in the next table, and graphs of the derivatives are shown after the table. There is considerable noise in the data, but the first derivative appears to reach a minimum near 27.47 mL. The curve through the data was drawn by hand. There is too much noise in the data to use the second derivative. The first derivative near the second end point is computed in the second table and plotted in the subsequent graph. The derivative reaches a minimum near 41.22 mL. The uncertainty in both end points is probably in the 0.1-mL decimal place.

The quotient of the two end points is $41.22/27.47 = 1.501$, which is 3:2, suggesting that the unknown is tribasic. We are probably observing the end points for the reactions $B + 2H^+ \rightarrow BH_2^{2+}$ and $B + 3H^+ \rightarrow BH_3^{3+}$. The molarity of the base from the first end point is

$$[B] = \frac{\frac{1}{2}(\text{mmol acid})}{100.0 \text{ mL}} = \frac{\frac{1}{2}(27.47 \text{ mL})(0.111\ 4 \text{ M})}{100.0 \text{ mL}} = 0.015\ 3_0 \text{ M}$$

The molarity from the second end point is

$$[B] = \frac{\frac{1}{3}(\text{mmol acid})}{100.0 \text{ mL}} = \frac{\frac{1}{3}(41.22 \text{ mL})(0.1114 \text{ M})}{100.0 \text{ mL}} = 0.0153_1 \text{ M}$$

First end point					
		First derivative		Second derivative	
Volume (mL)	pH	mL	$\frac{\Delta(\text{pH})}{\Delta(\text{mL})}$	mL	$\frac{\Delta[\Delta(\text{pH})/\Delta(\text{mL})]}{\Delta(\text{mL})}$
26.939	8.217				
27.013	8.149	26.9760	−0.9189		
27.067	8.096	27.0400	−0.9815	27.0080	−0.9775
27.114	8.050	27.0905	−0.9787	27.0653	0.0546
27.165	7.987	27.1395	−1.2353	27.1150	−5.2361
27.213	7.916	27.1890	−1.4792	27.1643	−4.9267
27.248	7.856	27.2305	−1.7143	27.2098	−5.6655
27.280	7.791	27.2640	−2.0312	27.2473	−9.4616
27.309	7.734	27.2945	−1.9655	27.2793	2.1552
27.338	7.666	27.3235	−2.3448	27.3090	−13.0797
27.362	7.603	27.3500	−2.6250	27.3368	−10.5725
27.386	7.538	27.3740	−2.7083	27.3620	−3.4722
27.406	7.485	27.3960	−2.6500	27.3850	2.6515
27.427	7.418	27.4165	−3.1905	27.4063	−26.3647
27.444	7.358	27.4355	−3.5294	27.4260	−17.8387
27.463	7.287	27.4535	−3.7368	27.4445	−11.5239
27.481	7.228	27.4720	−3.2778	27.4628	24.8143
27.501	7.158	27.4910	−3.5000	27.4815	−11.6959
27.517	7.103	27.5090	−3.4375	27.5000	3.4722
27.537	7.049	27.5270	−2.7000	27.5180	40.9722
27.558	6.982	27.5475	−3.1905	27.5373	−23.9257
27.579	6.920	27.5685	−2.9524	27.5580	11.3379
27.600	6.871	27.5895	−2.3333	27.5790	29.4785
27.622	6.825	27.6110	−2.0909	27.6003	11.2755
27.649	6.769	27.6355	−2.0741	27.6233	0.6871
27.675	6.717	27.6620	−2.0000	27.6488	2.7952
27.714	6.646	27.6945	−1.8205	27.6783	5.5227
27.747	6.594	27.7305	−1.5758	27.7125	6.7988
27.793	6.535	27.7700	−1.2826	27.7503	7.4215
27.846	6.470	27.8195	−1.2264	27.7948	1.1352
27.902	6.411	27.8740	−1.0536	27.8468	3.1714
27.969	6.347	27.9355	−0.9552	27.9048	1.5991

How can we deliver volumes up to 50 mL with a precision of 0.001 mL? Probably the best way is to measure the mass of titrant solution delivered in each addition. We can measure a 50-gram quantity of titrant to the 0.001-g decimal place, which provides the required precision. We would need to measure the density of titrant to 5 significant figures in order to convert the 5-significant-figure precision to 5-significant-figure accuracy.

Acid-Base Titrations

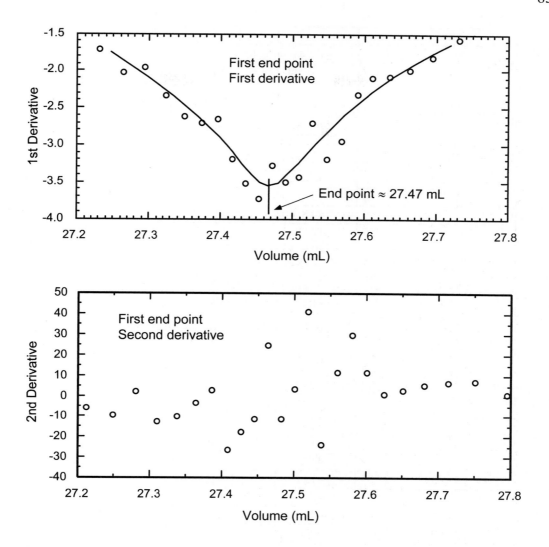

Second end point		First derivative	
Volume (mL)	pH	mL	$\frac{\Delta(pH)}{\Delta(mL)}$
40.168	3.877		
40.403	3.767	40.2855	−0.4681
40.498	3.728	40.4505	−0.4105
40.604	3.669	40.5510	−0.5566
40.680	3.618	40.6420	−0.6711
40.774	3.559	40.7270	−0.6277
40.854	3.51	40.8140	−0.6125
40.925	3.457	40.8895	−0.7465
40.994	3.407	40.9595	−0.7246
41.057	3.363	41.0255	−0.6984
41.114	3.317	41.0855	−0.8070
41.184	3.263	41.1490	−0.7714
41.254	3.21	41.2190	−0.7571
41.329	3.15	41.2915	−0.8000
41.406	3.093	41.3675	−0.7403
41.466	3.047	41.4360	−0.7667
41.542	2.999	41.5040	−0.6316
41.620	2.949	41.5810	−0.6410
41.717	2.887	41.6685	−0.6392
41.791	2.845	41.7540	−0.5676
41.905	2.795	41.8480	−0.4386
42.033	2.735	41.9690	−0.4688
42.351	2.617	42.1920	−0.3711
42.709	2.506	42.5300	−0.3101
43.192	2.401	42.9505	−0.2174
43.630	2.312	43.4110	−0.2032

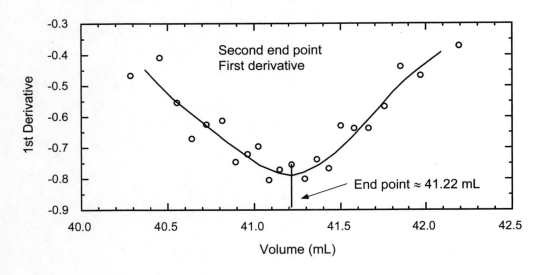

CHAPTER 11
POLYPROTIC ACIDS AND BASES

11-1. Point A is a solution of a weak acid, H_2A, which effectively behaves as a monoprotic acid whose pH is computed from the reaction $H_2A \rightleftharpoons HA^- + H^+$. Point B is a mixture of H_2A and HA^-, which is a buffer whose pH is computed from the Henderson-Hasselbalch equation $pH = pK_1 + \log([HA^-]/[H_2A])$. Point C is the first equivalence point—a solution of HA^- whose pH is given approximately by $\frac{1}{2}(pK_1 + pK_2)$. Point D is a mixture of HA^- and A^{2-}, which is a buffer whose pH is computed from another Henderson-Hasselbalch equation: $pH = pK_2 + \log([A^{2-}]/[HA^-])$. Point E is the second equivalence point, which is a solution of the weak base A^{2-} that can be treated as monobasic for computing the pH: $A^{2-} + H_2O \rightleftharpoons HA^- + OH^-$.

11-2. $HSO_4^- \xrightleftharpoons[]{K_{a2} = 1.03 \times 10^{-2}} SO_4^{2-} + H^+$ (K_{a2} comes from Appendix B)

$HC_2O_4^- + H_2O \xrightleftharpoons[]{K_{b2} = K_w/K_{a1} = 1.78 \times 10^{-13}} H_2C_2O_4 + OH^-$
(K_{a1} comes from Appendix B)

11-3. $K_{a1} = \dfrac{K_w}{K_{b3}} = 7.09 \times 10^{-3}$ $\quad K_{a2} = \dfrac{K_w}{K_{b2}} = 6.33 \times 10^{-8}$

$K_{a3} = \dfrac{K_w}{K_{b1}} = 4.2 \times 10^{-13}$

11-4. $H_3\overset{+}{N}$—CHR—CO_2^- pK_a applies to —$\overset{+}{N}H_3$ and —CO_2H. If substituent R has an acidic proton, there is also a pK_a for R.

11-5. $HN\text{(ring)}NH + H_2O \xrightleftharpoons[]{K_{b1} = K_w/K_{a2} = 5.38 \times 10^{-5}} HN\text{(ring)}NH_2^+ + OH^-$

$HN\text{(ring)}NH_2^+ + H_2O \xrightleftharpoons[]{K_{b2} = K_w/K_{a1} = 2.15 \times 10^{-9}} H_2\overset{+}{N}\text{(ring)}NH_2^+ + OH^-$

phthalate(CO_2^-, CO_2^-) $+ H_2O \xrightleftharpoons[]{K_{b1} = K_w/K_{a2} = 2.56 \times 10^{-9}}$ phthalate(CO_2H, CO_2^-)

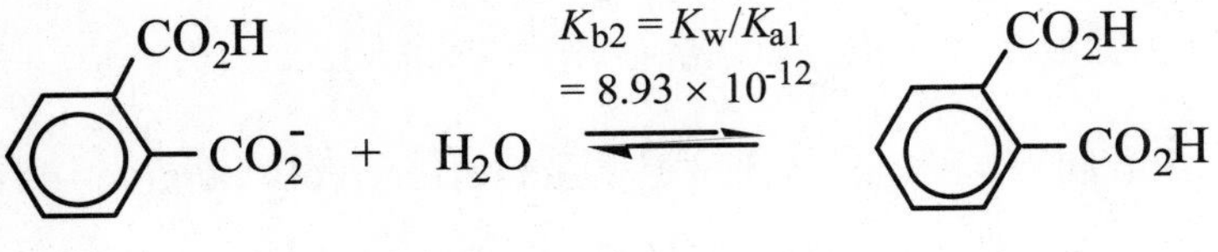

11-6.

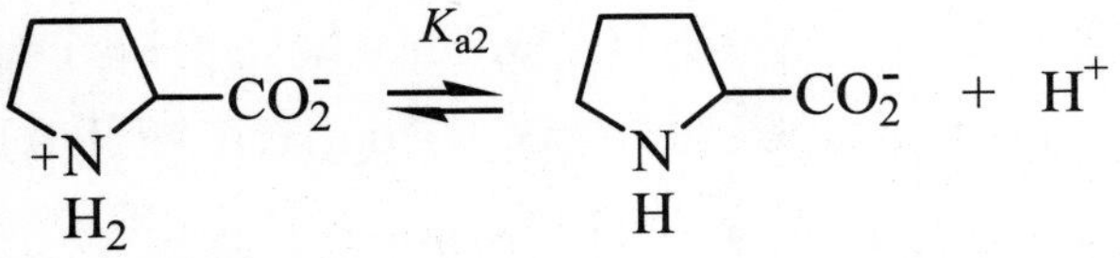

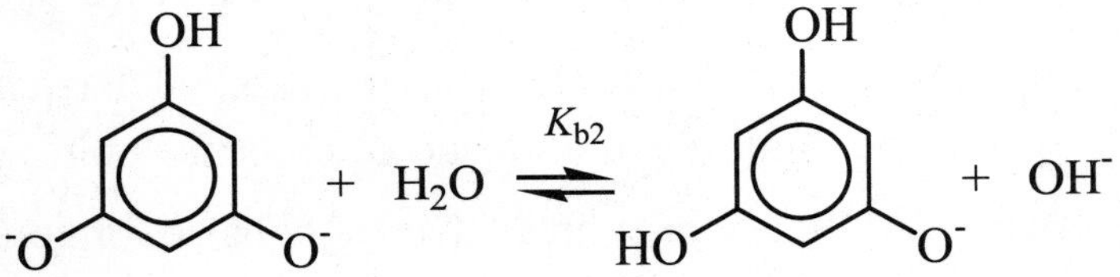

[structure: pyrrolidinium with CO$_2^-$] $\xrightleftharpoons{K_{a2}}$ [structure: pyrrolidine with CO$_2^-$] + H$^+$

[structure: 3,5-dioxidophenol] + H$_2$O $\xrightleftharpoons{K_{b2}}$ [structure: 3-hydroxy-5-oxidophenol] + OH$^-$

11-7. $K_{b1} = \dfrac{K_w}{K_{a3}} = 2.49 \times 10^{-8}$

$K_{b2} = \dfrac{K_w}{K_{a2}} = 5.78 \times 10^{-10}$

$K_{b3} = \dfrac{K_w}{K_{a1}} = 1.34 \times 10^{-11}$

11-8.

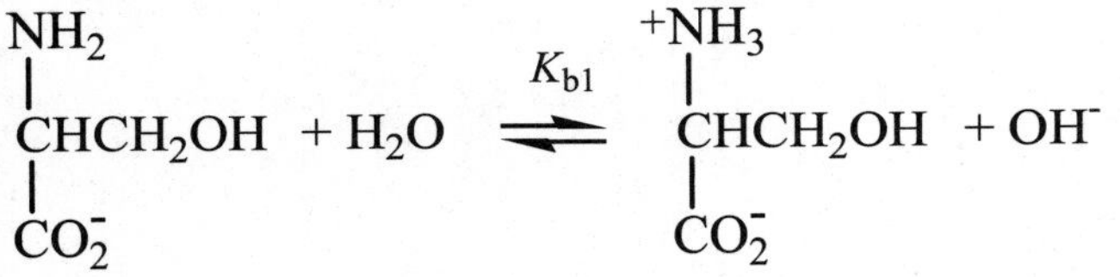

[serine anion NH$_2$–CH(CH$_2$OH)–CO$_2^-$] + H$_2$O $\xrightleftharpoons{K_{b1}}$ [$^+$NH$_3$–CH(CH$_2$OH)–CO$_2^-$] + OH$^-$ $K_{b1} = K_w/K_{a2} = 1.62 \times 10^{-5}$

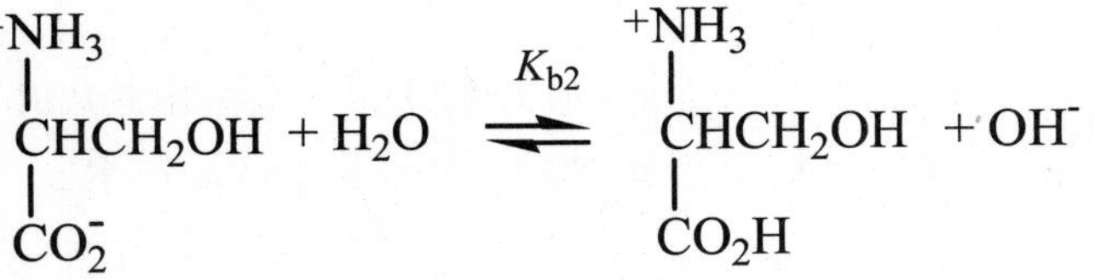

[$^+$NH$_3$–CH(CH$_2$OH)–CO$_2^-$] + H$_2$O $\xrightleftharpoons{K_{b2}}$ [$^+$NH$_3$–CH(CH$_2$OH)–CO$_2$H] + OH$^-$ $K_{b2} = K_w/K_{a1} = 1.54 \times 10^{-12}$

11-9. (a) We treat H$_2$M as a monoprotic acid:

H$_2$M $\rightleftharpoons$ H$^+$ + HM$^-$ $K_1 = 1.42 \times 10^{-3}$
F$-x$ x x

$\dfrac{x^2}{0.100 - x} = K_1 \Rightarrow x = 1.12 \times 10^{-2} \Rightarrow$ pH $= -\log x = 1.95$

[H$_2$M] $= 0.100 - x = 0.089$ M

[HM$^-$] $= x = 1.12 \times 10^{-2}$ M   [M^{2-}] $= \dfrac{[\text{HM}^-]K_2}{[\text{H}^+]} = 2.01 \times 10^{-6}$ M

(b) HM^- is the intermediate form of a diprotic acid. Therefore,

$$pH \approx \frac{1}{2}(pK_1 + pK_2) = \frac{1}{2}(2.847 + 5.696) = 4.27 \Rightarrow [H^+] = 5.3_5 \times 10^{-5}\ M$$

$$[HM^-] \approx 0.100\ M \qquad [H_2M] = \frac{[HM^-][H^+]}{K_1} = 3.8 \times 10^{-3}\ M$$

$$[M^{2-}] = \frac{K_2[HM^-]}{[H^+]} = 3.8 \times 10^{-3}\ M$$

(c) $M^{2-} + H_2O \rightleftharpoons HM^- + OH^- \qquad K_{b1} = K_w/K_{a2} = 4.98 \times 10^{-9}$
$\ F-x x x$

$$\frac{x^2}{0.100-x} = K_{b1} \Rightarrow x = 2.23 \times 10^{-5} \Rightarrow pH = -\log\frac{K_w}{x} = 9.35$$

$$[M^{2-}] = 0.100 - x = 0.100\ M \qquad [HM^-] = x = 2.23 \times 10^{-5}\ M$$

$$[H_2M] = \frac{[H^+][HM^-]}{K_1} = 7.04 \times 10^{-12}$$

11-10. (a) We treat B as if it were a monoprotic base:

$$B + H_2O \xrightleftharpoons[]{K_{b1} = 1.00 \times 10^{-5}} BH^+ + OH^- \qquad K_{a1} = K_w/K_{b2} = 1.00 \times 10^{-5}$$
$$0.100-x x x \qquad K_{a2} = K_w/K_{b1} = 1.00 \times 10^{-9}$$

$$\frac{x^2}{0.100-x} = K_{b1} \Rightarrow x = 9.95 \times 10^{-4} \Rightarrow [H^+] = \frac{K_w}{x} = 1.01 \times 10^{-11}\ M$$

$$\Rightarrow pH = 11.00$$

$$[BH^+] = x = 9.95 \times 10^{-4}\ M \qquad [B] = 0.100 - x = 0.099_0\ M$$

$$[BH_2^{2+}] = \frac{[BH^+][H^+]}{K_{a1}} = 1.00 \times 10^{-9}\ M$$

(b) BH^+ is the intermediate form of a diprotic acid with acid dissociation constants K_{a1} and K_{a2} computed in part (a).

$$pH \approx \frac{1}{2}(pK_{a1} + pK_{a2}) = \frac{1}{2}(5.00 + 9.00) = 7.00 \Rightarrow [H^+] = 1.0_0 \times 10^{-7}\ M$$

$$[BH^+] \approx 0.100\ M \qquad [BH_2^{2+}] = \frac{[BH^+][H^+]}{K_{a1}} = 1.0 \times 10^{-3}\ M$$

$$[B] = \frac{K_{a2}[BH^+]}{[H^+]} = 1.0 \times 10^{-3}\ M$$

(c) We treat BH_2^{2+} as a monoprotic acid:

$$BH_2^{2+} \rightleftharpoons BH^+ + H^+ \qquad K_{a1} = 1.00 \times 10^{-5}$$
$$0.100-x x x$$

$$\frac{x^2}{0.100-x} = K_{a1} \Rightarrow x = [H^+] = 9.95 \times 10^{-4}\ M \Rightarrow pH = 3.00$$

$$[BH^+] = x = 9.95 \times 10^{-4}\ M \qquad [BH_2^{2+}] = 0.100 - x = 0.099_0\ M$$

$$[B] = \frac{K_{a2}[BH^+]}{[H^+]} = 1.00 \times 10^{-9} \text{ M}$$

	pH	[B]	[BH$^+$]	[BH$_2^{2+}$]
0.100 M B	11.00	0.0990	9.95×10^{-4}	1.00×10^{-9}
0.100 M BH$^+$	7.00	1.0×10^{-3}	0.100	1.0×10^{-3}
0.100 M BH$_2^{2+}$	3.00	1.00×10^{-9}	9.95×10^{-4}	0.0990

11-11. HN⌬NH + H$_2$O ⇌ HN⌬NH$_2^+$ + OH$^-$
 $K_{b1} = K_w/K_{a2} = 5.38 \times 10^{-5}$

F − x, x, x

$$\frac{x^2}{0.300-x} = K_{b1} \Rightarrow x = 3.99 \times 10^{-3} \text{ M} \Rightarrow \text{pH} = -\log K_w/x = 11.60$$

$[B] = 0.300 - x = 0.296 \text{ M}$ $\quad [BH^+] = x = 3.99 \times 10^{-3} \text{ M}$

$$[BH_2^{2+}] = \frac{[BH^+][H^+]}{K_1} = 2.15 \times 10^{-9} \text{ M}$$

11-12. HN⌬NH$_2^+$ (which we designate BH$^+$) is the intermediate form of a diprotic acid with F = 0.150 M, $K_1 = 4.65 \times 10^{-6}$, and $K_2 = 1.86 \times 10^{-10}$.

$$\text{pH} \approx \tfrac{1}{2}(pK_1 + pK_2) = \tfrac{1}{2}(5.333 + 9.731) = 7.53 \Rightarrow [H^+] = 2.9_4 \times 10^{-8} \text{ M}$$

$[BH^+] \approx 0.150 \text{ M}$ $\quad [BH_2^{2+}] = \dfrac{[BH^+][H^+]}{K_1} = 9.4_8 \times 10^{-4} \text{ M}$

$[B] = \dfrac{K_2[BH^+]}{[H^+]} = 9.4_9 \times 10^{-4} \text{ M}$

In this problem, we find that $[BH_2^{2+}] + [B] = 0.001\,9$ M. Therefore, a better approximation for [BH$^+$] would be $[BH^+] \approx 0.150 - 0.0019 = 0.148$ M. We could recompute the concentrations in this problem, using Equation 11-11 with F = 0.148 M to obtain a slightly better set of answers.

11-13.
$$\begin{array}{c} \overset{+}{NH_3} \\ | \\ CH-CH_2CH_2\overset{O}{\overset{\|}{C}}NH_2 \\ | \\ CO_2^- \end{array}$$

Glutamine is the intermediate form of a diprotic acid, so pH $\approx \tfrac{1}{2}(pK_1 + pK_2)$ $= \tfrac{1}{2}(2.19 + 9.00) = 5.60.$

Polyprotic Acids and Bases

11-14. (a) Diprotic system:

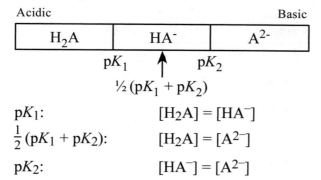

pK_1:	$[H_2A] = [HA^-]$
$\frac{1}{2}(pK_1 + pK_2)$:	$[H_2A] = [A^{2-}]$
pK_2:	$[HA^-] = [A^{2-}]$

(b) Monoprotic system:

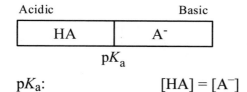

pK_a: $[HA] = [A^-]$

Triprotic system:

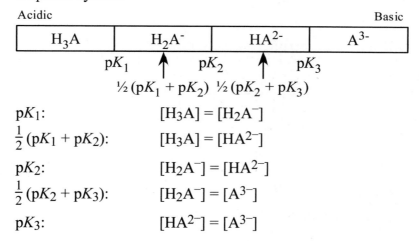

pK_1:	$[H_3A] = [H_2A^-]$
$\frac{1}{2}(pK_1 + pK_2)$:	$[H_3A] = [HA^{2-}]$
pK_2:	$[H_2A^-] = [HA^{2-}]$
$\frac{1}{2}(pK_2 + pK_3)$:	$[H_2A^-] = [A^{3-}]$
pK_3:	$[HA^{2-}] = [A^{3-}]$

11-15. (a) At pH = pK_a, the concentrations of HA and A^- are equal. At a pH below pK_a, HA is the dominant form.

(b) At a pH above pK_a, A^- is the dominant form.

(c) $\text{pH} = pK_a + \dfrac{[A^-]}{[HA]}$

$7.00 = 7.00 + \log\dfrac{[A^-]}{[HA]} \;\Rightarrow\; [A^-]/[HA] = 1.0$

$6.00 = 7.00 + \log\dfrac{[A^-]}{[HA]} \;\Rightarrow\; [A^-]/[HA] = 0.10$

11-16. (a) 4.00 (b) 8.00 (c) H_2A (d) HA^- (e) A^{2-}

11-17.

| H_3PO_4 | $H_2PO_4^-$ | HPO_4^{2-} | PO_4^{3-} |

pH: 2.15 ↑ 7.20 ↑ 12.38
          4.68   9.78

pH	Dominant form	Second most prevalent form
2	H_3PO_4	$H_2PO_4^-$
3, 4	$H_2PO_4^-$	H_3PO_4
5, 6, 7	$H_2PO_4^-$	HPO_4^{2-}
8, 9	HPO_4^{2-}	$H_2PO_4^-$
10, 11, 12	HPO_4^{2-}	PO_4^{3-}
13	PO_4^{3-}	HPO_4^{2-}

11-18. (a) 9.00 (b) 9.00 (c) BH^+

(d) $12.00 = 9.00 + \log \frac{[B]}{[BH^+]} \Rightarrow [B]/[BH^+] = 1.0 \times 10^3$

11-19. (a) $K_{a1} = K_w/K_{b2} \Rightarrow pK_{a1} = pK_w - pK_{b2} = 14.00 - 7.15 = 6.85$

$K_{a2} = K_w/K_{b1} \Rightarrow pK_{a2} = pK_w - pK_{b1} = 14.00 - 4.07 = 9.93$

| BH_2^{2+} | BH^+ | B |

pH: 6.85 ↑ 9.93
          8.39

(b) $[BH^+] = [B]$ at pH 9.93

(c) $[BH^+] = [BH_2^{2+}]$ at pH 6.85

(d)
pH	Dominant form	Second most prevalent form
4, 5, 6	BH_2^{2+}	BH^+
7, 8	BH^+	BH_2^{2+}
9	BH^+	B
10	B	BH^+

(e) At pH 12.00, B is dominant and BH^+ is the second most abundant species. There is a negligible amount of BH_2^{2+}. The Henderson-Hasselbalch equation for B and BH^+ is

$$pH = pK_{a2} + \log \frac{[B]}{[BH^+]}$$

Polyprotic Acids and Bases

$$12.00 = 9.93 + \log \frac{[B]}{[BH^+]} \Rightarrow [B]/[BH^+] = 1.2 \times 10^2$$

(f) At pH 2.00, BH_2^{2+} is dominant and BH^+ is the second most abundant species. There is a negligible amount of B. The Henderson-Hasselbalch equation for BH_2^{2+} and BH^+ is

$$pH = pK_{a1} + \log \frac{[BH^+]}{[BH_2^{2+}]}$$

$$2.00 = 6.85 + \log \frac{[BH^+]}{[BH_2^{2+}]} \Rightarrow [BH^+]/[BH_2^{2+}] = 1.4 \times 10^{-5}$$

$$\Rightarrow [BH_2^{2+}]/[BH^+] = 7.1 \times 10^4$$

11-20.

	pH 9.00	pH 10.00
Predominant form:	$^+NH_3$ on ^-O_2C–CH–(CH$_2$)–CO_2^-	NH_2 on ^-O_2C–CH–(CH$_2$)–CO_2^-
Secondary form:	NH_2 on ^-O_2C–CH–(CH$_2$)–CO_2^-	$^+NH_3$ on ^-O_2C–CH–(CH$_2$)–CO_2^-

Predominant form:	NH_2; ^-O_2C–CH–(C$_6$H$_4$)–OH	NH_2; ^-O_2C–CH–(C$_6$H$_4$)–OH
Secondary form:	$^+NH_3$; ^-O_2C–CH–(C$_6$H$_4$)–OH	NH_2; ^-O_2C–CH–(C$_6$H$_4$)–O$^-$

11-21. (a) Calling the three forms of glutamine H_2G^+, HG, and G^-, the form shown is G^-, which is the fully basic form. Treating G^- as monobasic, we write

$$G^- + H_2O \rightleftharpoons HG + OH^- \qquad K_{b1} = K_w/K_{a2} = 1.0_0 \times 10^{-5}$$

$$F - x \qquad\qquad x \qquad x$$

$$\frac{x^2}{0.10 - x} = K_{b1} \Rightarrow x = 9.9_5 \times 10^{-4}\ M \Rightarrow pH = -\log K_w/x = 11.00$$

(b) Calling the four forms of arginine H_3A^{2+}, H_2A^+, HA, and A^-, the form shown is HA.

$$pH \approx \tfrac{1}{2}(pK_2 + pK_3) = \tfrac{1}{2}(8.991 + 12.1) = 10.5$$

11-22. Pyridoxal-5-phosphate is a tetraprotic system with pK_a values of 1.4, 3.44, 6.01, and 8.45. At pH 7.00, the predominant species has lost 3 of the 4 protons.

11-23. For H_3Arg^{2+}, $pK_1 = 1.823$
$pK_2 = 8.991$
$pK_3 = 12.1$

H_2Arg^+ found in Arginine · HCl

$$pH \approx \tfrac{1}{2}(pK_1 + pK_2) = 5.41 \Rightarrow [H^+] = 3.9 \times 10^{-6} \text{ M}$$

$$\frac{[H_2Arg^+][H^+]}{[H_3Arg^{2+}]} = K_1 \Rightarrow [H_3Arg^{2+}] = \frac{(0.050)(3.9 \times 10^{-6})}{10^{-1.823}} = 1.3 \times 10^{-5} \text{ M}$$

$$\frac{[HArg][H^+]}{[H_2Arg^+]} = K_2 \Rightarrow [HArg] = \frac{(10^{-8.991})(0.050)}{3.9 \times 10^{-6}} = 1.3 \times 10^{-5} \text{ M}$$

$$\frac{[Arg^-][H^+]}{[HArg]} = K_3 \Rightarrow [Arg^-] = \frac{(10^{-12.1})(1.3 \times 10^{-5})}{3.9 \times 10^{-6}} = 3 \times 10^{-12} \text{ M}$$

11-24. $H_3Cit \xrightleftharpoons[]{pK_1 = 3.128} H_2Cit^- \xrightleftharpoons[]{pK_2 = 4.761} HCit^{2-} \xrightleftharpoons[]{pK_3 = 6.396} Cit^{3-}$

At pH 5.00, $HCit^{2-}$ is dominant.

11-25. $\underbrace{(100.0 \text{ mL})(0.100 \text{ M})}_{\text{mmol } H_2A} = \underbrace{V_{e1}(1.00 \text{ M})}_{\text{mmol } OH^-} \Rightarrow V_{e1} = 10.0 \text{ mL}$

$V_{e2} = 2V_{e1} = 20.0 \text{ mL}$

V_b:	0	5.0	10.0	15.0	20.0	22.0
pH:	2.51	4.00	6.00	8.00	10.46	12.21

Polyprotic Acids and Bases

0 mL: This is a solution of H_2A, which we treat as a monoprotic acid.

$$H_2A \rightleftharpoons HA^- + H^+ \qquad \frac{x^2}{0.100-x} = K_1 = 10^{-4.00}$$

$$0.100-x \quad x \quad x$$

$$\Rightarrow x = 3.11 \times 10^{-3}\ M \Rightarrow pH = 2.51$$

5.0 mL: Halfway to the first equivalence point, we have an equimolar mixture of H_2A and HA^-, (Aha! A buffer!) with $pH = pK_1 = 4.00$.

10.0 mL: The first equivalence point gives a solution of HA^-, the intermediate form of a diprotic acid, with $pH \approx \frac{1}{2}(pK_1 + pK_2) = 6.00$.

15.0 mL: Halfway to the second equivalence point, we have an equimolar mixture of HA^- and A^{2-}, with $pH = pK_2 = 8.00$.

20.0 mL: At the second equivalence point, we have a solution of A^{2-}, which we treat as monobasic with a formal concentration of $F' = 0.083\ 3\ M$ because the original 100 mL solution has been diluted to 120 mL.

$$A^{2-} + H_2O \xrightleftharpoons{K_{b1}} HA^- + OH^-$$

$$F' - x \qquad\qquad\qquad x \quad\ \ x$$

$$F' = \frac{(100\ mL)(0.100\ M)}{120\ mL} = 0.083\ 3\ M$$

$$\frac{x^2}{0.083\ 3 - x} = K_{b1} = \frac{K_w}{K_2} \Rightarrow x = 2.88 \times 10^{-4}\ M$$

$$\Rightarrow pH = -\log \frac{K_w}{x} = 10.46$$

22.0 mL: Now there are 2.0 mL of excess OH^-.

$$[OH^-] = \frac{(2.0\ mL)(1.00\ M)}{122.0\ mL} = 1.64 \times 10^{-2}\ M \Rightarrow pH = 12.21$$

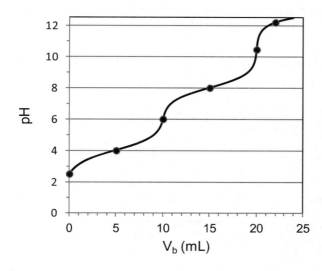

11-26. $\underbrace{(100.0 \text{ mL})(0.100 \text{ M})}_{\text{mmol H}_2\text{A}} = \underbrace{V_{e1}(1.00 \text{ M})}_{\text{mmol OH}^-} \Rightarrow V_{e1} = 10.0 \text{ mL}$

$V_{e2} = 2V_{e1} = 20.0 \text{ mL}$

K_{a1} (for BH_2^{2+}) $= K_w/K_{b2} = 10^{-6.00}$ $pK_{a1} = 6.00$

K_{a2} (for BH_2^{2+}) $= K_w/K_{b1} = 10^{-10.00}$ $pK_{a2} = 10.00$

V_a	0	5	10	15	20	22
pH	11.49	10.00	8.00	6.00	3.54	1.79

0 mL: This is a solution of B, which we treat as monobasic.

$$\text{B} + \text{H}_2\text{O} \underset{}{\overset{K_{b1}}{\rightleftharpoons}} \text{BH}^+ + \text{OH}^-$$
$$0.100-x \qquad\qquad x \qquad x$$

$\dfrac{x^2}{0.100-x} = 10^{-4.00} \Rightarrow x = 3.11 \times 10^{-3} \text{ M}$

$\Rightarrow \text{pH} = -\log\dfrac{K_w}{x} = 11.49$

5.0 mL: Halfway to the first equivalence point, we have an equimolar mixture of B and BH^+, for which pH $= pK_{a2} = 10.00$.

10.0 mL: At the first equivalence point, B has been converted to BH^+, the intermediate form of a diprotic system.

$\text{pH} \approx \dfrac{1}{2}(pK_{a1} + pK_{a2}) = \dfrac{1}{2}(6.00 + 10.00) = 8.00$

15.0 mL: Halfway to the second equivalence point, we have an equimolar mixture of BH^+ and BH_2^{2+}, for which pH $= pK_{a1} = 6.00$.

20.0 mL: At the second equivalence point, B has been converted to BH_2^{2+}, which we treat as a monoprotic acid. The formal concentration is $F' = 0.083\ 3$ M because the original 100-mL solution has been diluted to 120 mL.

$$BH_2^{2+} \overset{K_{a1}}{\rightleftharpoons} BH^+ + H^+$$
$$F'-x \qquad\qquad x \qquad x$$

$F' = \dfrac{(100.0 \text{ mL})(0.100 \text{ M})}{120.0 \text{ mL}} = 0.083\ 3 \text{ M}$

$\dfrac{x^2}{0.083\ 3 - x} = K_{a1} = 10^{-6.00} \Rightarrow x = 2.88 \times 10^{-4} \text{ M} \Rightarrow \text{pH} = 3.54$

22.0 mL: At this point, the pH is mainly governed by the 2.0 mL of strong acid added beyond the second equivalence point.

$[H^+] = \dfrac{(2.0 \text{ mL})(1.00 \text{ M})}{122.0 \text{ mL}} = 1.64 \times 10^{-2} \text{ M} \Rightarrow \text{pH} = 1.79$

Polyprotic Acids and Bases

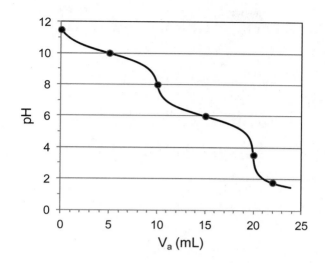

11-27. (a) phenolphthalein (colorless → red)

(b) *p*-nitrophenol (colorless → yellow)

(c) Here are two choices: bromothymol blue (yellow → green) or bromocresol purple (yellow → yellow + purple)

(d) thymolphthalein (colorless → blue)

11-28. Titration reactions:

$$HN\bigcirc NH + H^+ \longrightarrow H_2\overset{+}{N}\bigcirc NH \qquad V_{e1} = 40.0 \text{ mL}$$

$$H_2\overset{+}{N}\bigcirc NH + H^+ \longrightarrow H_2\overset{+}{N}\bigcirc \overset{+}{N}H_2 \qquad V_{e2} = 80.0 \text{ mL}$$

$$K_{b1} = \frac{K_w}{K_{a2}} = \frac{10^{-14.00}}{10^{-9.731}} = 5.38 \times 10^{-5} \qquad K_{b2} = \frac{K_w}{K_{a1}} = \frac{10^{-14.00}}{10^{-5.333}} = 2.15 \times 10^{-9}$$

0 mL: $\quad B + H_2O \overset{K_{b1}}{\rightleftharpoons} BH^+ + OH^- \qquad \frac{x^2}{0.100-x} = K_{b1}$

$\qquad \quad 0.100-x \qquad \qquad x \qquad x \qquad \Rightarrow x = 2.29 \times 10^{-3} \text{ M}$

$\qquad \qquad \qquad \qquad \qquad \qquad \qquad \qquad \Rightarrow \text{pH} = 11.36$

20.0 mL: Halfway to V_{e1}, there is an equimolar mixture of B and BH$^+$:
$\qquad \text{pH} = \text{p}K_2 = 9.73.$

40.0 mL: At V_{e1}, B has been converted to BH$^+$, the intermediate form of a diprotic acid: $\text{pH} \approx \frac{1}{2}(\text{p}K_1 + \text{p}K_2) = \frac{1}{2}(5.333 + 9.731) = 7.53.$

60.0 mL: Halfway to V_{e1}, there is an equimolar mixture of BH^+ and BH_2^{2+}:

$$pH = pK_1 = 5.33.$$

80.0 mL: B has been converted to BH_2^{2+} at a formal concentration of

$$\frac{(40.0 \text{ mL})(0.100 \text{ M})}{120.0 \text{ mL}} = 0.033\,3 \text{ M}$$

$$BH_2^{2+} \underset{}{\overset{K_{a1}}{\rightleftharpoons}} BH^+ + H^+ \qquad \frac{x^2}{0.033\,3-x} = K_{a1} = 10^{-5.333}$$

$$0.033\,3-x \qquad x \qquad x \qquad \Rightarrow x = 3.91 \times 10^{-4} \text{ M} \Rightarrow pH = 3.41$$

100.0 mL: Excess $[H^+] = \dfrac{(20.0 \text{ mL})(0.100 \text{ M})}{140.0 \text{ mL}} \Rightarrow pH = 1.85$

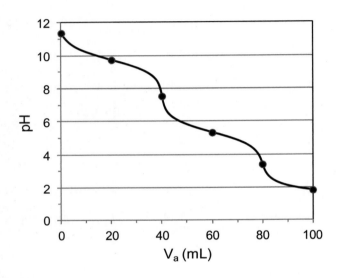

11-29. The titration reactions between the triprotic acid and the strong base are

$H_3PO_4 + OH^- \rightarrow H_2PO_4^- + H_2O$

$(25.0 \text{ mL})(0.0400 \text{ M}) = V_{e1}(0.0500 \text{ M}) \Rightarrow V_{e1} = 20.0 \text{ mL}$

$H_2PO_4^- + OH^- \rightarrow HPO_4^{2-} + H_2O \qquad V_{e2} = 2V_{e1} = 40.0 \text{ mL}$

$HPO_4^{2-} + OH^- \rightarrow PO_4^{3-} + H_2O \qquad V_{e3} = 3V_{e1} = 60.0 \text{ mL}$

The dominant species diagram for phosphoric acid looks like this:

H_3PO_4	$H_2PO_4^-$	HPO_4^{2-}	PO_4^{3-}
pH: 2.15	7.20	12.38	
pK$_1$ ↑ 4.67	pK$_2$ ↑ 9.79	pK$_3$	

0 mL: We have a solution of 0.040 0 M H_3PO_4 which we treat as monoprotic:

$$H_3PO_4 \underset{}{\overset{K_1}{\rightleftharpoons}} H_2PO_4^- + H^+ \qquad \frac{x^2}{0.040\,0-x} = K_1$$
$$\begin{array}{ccc} 0.040\,0-x & x & x \end{array}$$
$$\Rightarrow x = 0.013\,7\text{ M} \Rightarrow \text{pH} = 1.86$$

10.0 mL: Halfway to the first equivalence point, there is a 1:1 mixture of H_3PO_4 and $H_2PO_4^-$, so pH = pK_1 = 2.15.

20.0 mL: At the first equivalence point, H_3PO_4 has been converted to $H_2PO_4^-$, the first intermediate form of the polyprotic acid. Therefore, pH ≈ $\frac{1}{2}(pK_1 + pK_2) = \frac{1}{2}(2.15 + 7.20) = 4.68$.

30.0 mL: Halfway to the second equivalence point, there is a 1:1 mixture of $H_2PO_4^-$ and HPO_4^{2-}, so pH = pK_2 = 7.20.

40.0 mL: At the second equivalence point, H_3PO_4 has been converted to HPO_4^{2-}, the second intermediate form of the polyprotic acid. Therefore, pH ≈ $\frac{1}{2}(pK_2 + pK_3) = \frac{1}{2}(7.20 + 12.38) = 9.79$.

42.0 mL: We have passed the second equivalence point by 2.0 mL, which is 1/10 of the way from the second to the third equivalence point. We have converted 1/10 of HPO_4^{2-} into PO_4^{3-}, leaving 9/10 in the form HPO_4^{2-}. We have a buffer made of HPO_4^{2-} and PO_4^{3-}:

$$\text{pH} = pK_3 + \log\frac{[PO_4^{3-}]}{[HPO_4^{2-}]} = 12.38 + \log\frac{[1/10]}{[9/10]} = 11.43$$

The six points in the graph below are the ones we just calculated. The solid line was computed by spreadsheet, using exact equations for the titration. These equations do not make the simplifying approximations that we have been using. The approximate point we computed at 10 mL (halfway to V_{e1}) has a small error because H_3PO_4 is a relatively strong acid. Our approximations underestimated the fraction that is dissociated to $H_2PO_4^-$.

There is a more serious error for the pH computed at 42 mL, which is beyond V_{e2}. HPO_4^{2-} is such a weak acid (pK_3 = 12.38) that it hardly reacts with OH^-. It takes a great deal of 0.05 M OH^- added to the existing volume of solution to raise the pH toward 12. For example, if we added 2 mL of 0.05 M OH^- to 65 mL of plain water, the pH would be 11.17. Yet we computed that the pH at 42 mL in the titration would be 11.43. Our approximate methods are poor in this example beyond the second equivalence point.

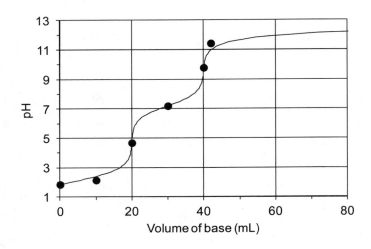

11-30.

(25.0 mL)(0.0500 M) = V_{e1}(0.0500 M) ⇒ V_{e1} = 25.0 mL

V_{e2} = 2V_{e1} = 50.0 mL

The equilibria of the triprotic acid H_3His^{2+} can be written as

$H_3His^{2+} \xrightleftharpoons{K_1} H_2His^+ \xrightleftharpoons{K_2} HHis \xrightleftharpoons{K_3} His^-$

0 mL: HHis is the second intermediate form derived from H_3His^{2+}.

$$pH \approx \tfrac{1}{2}(pK_2 + pK_3) = \tfrac{1}{2}(5.97 + 9.28) = 7.62$$

12.5 mL: Halfway to V_{e1}, there is an equimolar mixture of HHis and H_2His^+, so pH = pK_2 = 5.97.

25.0 mL: HHis has been converted to H_2His^+, which is the first intermediate form derived from H_3His^{2+}, so

$$pH \approx \tfrac{1}{2}(pK_1 + pK_2) = \tfrac{1}{2}(1.6 + 5.97) = 3.8$$

50.0 mL: Histidine has been converted to H_3His^{2+} at a formal concentration

$$F = \frac{(25.0 \text{ mL})(0.0500 \text{ M})}{75.0 \text{ mL}} = 0.0167 \text{ M}$$

$$\begin{array}{cccc} H_3His^{2+} & \rightleftharpoons & H_2His^+ + H^+ & K_1 = 0.02_5 \\ 0.0167 - x & & x \quad\quad x & \end{array}$$

$$\frac{x^2}{0.0167 - x} = K_1 \Rightarrow x = 0.01_{15} \text{ M} \Rightarrow \text{pH} = 1.9$$

11-31. (a) $\frac{1}{2}V_{e1}$: pH = pK_1 = 2.56

V_{e1}: pH ≈ $\frac{1}{2}(pK_1 + pK_2)$ = 3.46

$\frac{3}{2}V_{e1}$: pH = pK_2 = 4.37

V_{e2}: We need to estimate the formal concentration of A^{2-} at this point. Because 1 g H_2A ≈ 7.$_{57}$ mmol, we need ~15 mmol NaOH to reach V_{e2}. This much NaOH is contained in a volume of (15 mmol) / (0.0943$_2$ M) ≈ 160 mL.

$$[A^{2-}] \approx \frac{(100 \text{ mL})[(10 \text{ g/L})/(132.07 \text{ g/mol})]}{260 \text{ mL}} = 0.029 \text{ M}$$

$$\begin{array}{cccc} A^{2-} + H_2O & \rightleftharpoons & HA^- + OH^- & K_{b1} = \dfrac{K_w}{K_{a2}} = \dfrac{K_w}{10^{-4.37}} \\ 0.029 - x & & x \quad\quad x & \end{array}$$

$$\frac{x^2}{0.029 - x} = K_{b1} \Rightarrow x = 2.6 \times 10^{-6} \text{ M} \Rightarrow \text{pH} \approx 8.42$$

1.05 V_{e2}: The volume of NaOH is (1.05)(160 mL) = 168 mL, and the total volume of solution is 268 mL. The concentration of excess OH^- is (8/268)(0.0943$_2$ M) = 0.002 8$_2$ M.

$$[H^+] = K_w/[OH^-] = \frac{1.0 \times 10^{-14}}{0.002 \, 8_2}$$

$$= 3.55 \times 10^{-12} \text{ M} \Rightarrow \text{pH} = 11.45$$

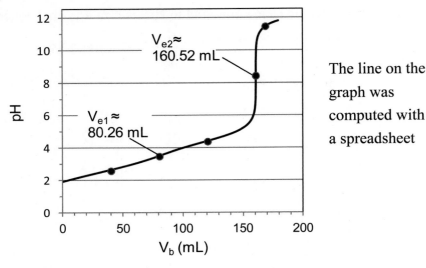

The line on the graph was computed with a spreadsheet

(b) Second (There is little break in the curve near the first equivalence point.)

(c) Thymolphthalein—first trace of blue

11-32. (a) From Equation C, $[CO_3^{2-}] = \dfrac{K_{a2}[HCO_3^-]}{[H^+]}$ (F)

From Equation B, $[HCO_3^-] = \dfrac{K_{a1}[CO_2(aq)]}{[H^+]}$ (G)

Substituting $[HCO_3^-]$ from Equation G into Equation F gives

$[CO_3^{2-}] = \dfrac{K_{a2} K_{a1}[CO_2(aq)]}{[H^+]^2}$ (H)

Substituting for $[CO_2(aq)]$ from Equation A into Equation H gives

$[CO_3^{2-}] = \dfrac{K_{a2} K_{a1} K_H P_{CO_2}}{[H^+]^2}$ (I)

(b) For $P_{CO_2} = 800$ μbar, pH = 7.8, 0°C, we find

$[CO_3^{2-}] = \dfrac{K_{a2} K_{a1} K_H P_{CO_2}}{[H^+]^2} =$

$\dfrac{10^{-9.3762}\ \text{mol kg}^{-1}\ 10^{-6.1004}\ \text{mol kg}^{-1}\ 10^{-1.2073}\ \text{mol kg}^{-1}\ \text{bar}^{-1}\ (800 \times 10^{-6}\ \text{bar})}{[10^{-7.8}\ \text{mol kg}^{-1}]^2}$

$= 6.6 \times 10^{-5}$ mol kg^{-1}

For $P_{CO_2} = 800$ μbar, pH = 7.8, 30°C, we find

$[CO_3^{2-}] = \dfrac{K_{a2} K_{a1} K_H P_{CO_2}}{[H^+]^2} =$

$\dfrac{10^{-8.8324}\ \text{mol kg}^{-1}\ 10^{-5.8008}\ \text{mol kg}^{-1}\ 10^{-1.6048}\ \text{mol kg}^{-1}\ \text{bar}^{-1}\ (800 \times 10^{-6}\ \text{bar})}{[10^{-7.8}\ \text{mol kg}^{-1}]^2}$

$= 1.8 \times 10^{-4}$ mol kg^{-1}

(c) The equilibrium expressions for aragonite and calcite are

$$CaCO_3(s, aragonite) \rightleftharpoons Ca^{2+} + CO_3^{2-} \quad (D)$$

$$K_{sp}^{arg} = [Ca^{2+}][CO_3^{2-}] = 10^{-6.1113} \text{ mol}^2 \text{ kg}^{-2} \text{ at } 0°C$$

$$= 10^{-6.1391} \text{ mol}^2 \text{ kg}^{-2} \text{ at } 30°C$$

$$CaCO_3(s, calcite) \rightleftharpoons Ca^{2+} + CO_3^{2-} \quad (E)$$

$$K_{sp}^{cal} = [Ca^{2+}][CO_3^{2-}] = 10^{-6.3652} \text{ mol}^2 \text{ kg}^{-2} \text{ at } 0°C$$

$$= 10^{-6.3713} \text{ mol}^2 \text{ kg}^{-2} \text{ at } 30°C$$

The reaction quotient at 0°C is

$$[Ca^{2+}][CO_3^{2-}]$$
$$= [0.010 \text{ mol kg}^{-1}][6.6 \times 10^{-5} \text{ mol kg}^{-1}] = 6.6 \times 10^{-7} \text{ mol}^2 \text{ kg}^{-2}$$
$$= 10^{-6.18} \text{ mol}^2 \text{ kg}^{-2} < 10^{-6.1113} \text{ mol}^2 \text{ kg}^{-2}$$

so aragonite will dissolve.

But $10^{-6.18}$ mol^2 kg^{-2} > $10^{-6.3652}$ mol^2 kg^{-2}, so calcite will not dissolve.

The reaction quotient at 30°C is

$$[Ca^{2+}][CO_3^{2-}]$$
$$= [0.010 \text{ mol kg}^{-1}][1.84 \times 10^{-4} \text{ mol kg}^{-1}] = 1.84 \times 10^{-6} \text{ mol}^2 \text{ kg}^{-2}$$
$$= 10^{-5.74} \text{ mol}^2 \text{ kg}^{-2} > 10^{-6.1113} \text{ mol}^2 \text{ kg}^{-2}$$

so neither aragonite nor calcite dissolve.

CHAPTER 12
A DEEPER LOOK AT CHEMICAL EQUILIBRIUM

12-1. The ionic atmosphere is the region of solution containing excess charge around any ion. For example, a Na$^+$ ion is surrounded by solution that is richer in anions than in cations.

12-2. As the ionic strength increases, the charges of the ionic atmospheres increase and the net ionic attractions decrease.

12-3. As ionic strength increases from zero, the activity coefficients of $CH_3CO_2^-$ and H$^+$ decrease, so their concentrations must increase if the equilibrium constant is to be really constant. The activity coefficient of the neutral species CH_3CO_2H is relatively insensitive to ionic strength. Beyond an ionic strength of ~0.6 M, the activity coefficients of both ions begin to rise, so the concentrations decrease and quotient $[CH_3CO_2^-][H^+]/[CH_3CO_2H]$ decreases.

12-4. (a) true   (b) true   (c) true

12-5. (a) Mg^{2+} has a smaller ionic diameter than Ba^{2+}, but equal charge, so Mg^{2+} binds water molecules more strongly and has a larger hydrated diameter.

(b) All of these ions are in the same row of the periodic table, so they have approximately similar atomic size. As the charge on the ion decreases, water is bound less strongly and the hydrated diameter decreases.

(c) H$_3$O$^+$ is tightly hydrogen bonded to three H$_2$O molecules and more loosely attracted to another by ion-dipole attraction. The overall size of the tightly bound cluster is considerably larger than H$_3$O$^+$.

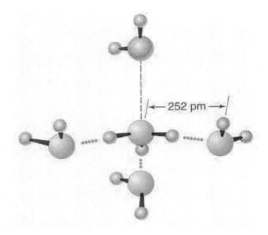

A Deeper Look at Chemical Equilibrium

12-6. A solution must have no net charge. The sum of positive charges in a solution must balance the sum of negative charges.

12-7. The number of atoms of an element (or groups of atoms) in all of its forms in a solution is equal to the number of atoms of that element (or group of atoms) delivered to the solution.

12-8. As pH is lowered, [H$^+$] increases. H$^+$ reacts with basic anions to increase the solubility of their salts. By promoting solubility of minerals such as galena and cerussite, the concentration of dissolved lead in the environment is increased.

Galena: $PbS(s) + H^+ \rightleftharpoons Pb^{2+} + HS^-$
Cerussite: $PbCO_3(s) + H^+ \rightleftharpoons Pb^{2+} + HCO_3^-$

Kaolinite and bauxite both have OH$^-$ ions that can react with H$^+$. The silicate groups in kaolinite and oxide in bauxite also react with H$^+$.

12-9. (a) $\mu = \frac{1}{2}([K^+] \cdot 1^2 + [NO_3^-] \cdot (-1)^2)$

$= \frac{1}{2}([0.2 \text{ mM}] \cdot 1^2 + [0.2 \text{ mM}] \cdot (-1)^2) = 0.2$ mM

(b) $\mu = \frac{1}{2}([Cs^+] \cdot 1^2 + [CrO_4^{2-}] \cdot (-2)^2) = \frac{1}{2}([0.4] \cdot 1 + [0.2] \cdot 4) = 0.6$ mM

(c) $\mu = \frac{1}{2}([Mg^{2+}] \cdot 2^2 + [Cl^-] \cdot (-1)^2 + [Al^{3+}] \cdot 3^2)$

$= \frac{1}{2}([0.2] \cdot 4 + [\ \ 0.4\ \ +\ \ 0.9\ \ ] \cdot 1 + [0.3] \cdot 9) = 2.4$ mM
  ↑   ↑
From MgCl$_2$ From AlCl$_3$

12-10. (a) 0.660   (b) 0.54   (c) 0.18 (Eu^{3+} is a lanthanide ion)   (d) 0.83

12-11. For 0.005 0 M (CH$_3$CH$_2$CH$_2$)$_4$N$^+$Br$^-$ plus 0.005 0 M (CH$_3$)$_4$N$^+$Cl$^-$, μ = 0.010 M. The size of the ion (CH$_3$CH$_2$CH$_2$)$_4$N$^+$ is 800 pm.

At μ = 0.01 M, γ = 0.912 for an ion of charge ±1 with α = 800 pm.
$\mathcal{A}$ = (0.005 0)(0.912) = 0.004 6

12-12. (a) The ionic strength 0.030 M is halfway between the values 0.01 and 0.05 M. Therefore, the activity coefficient will be halfway between the tabulated values:

$$\mu = \tfrac{1}{2}(0.914 + 0.86) = 0.88_7$$

(b) $\mu = 0.01$ 0.042   0.05

H^+: $\gamma = 0.914$?   0.86

$$\frac{\text{unknown } \gamma \text{ interval}}{\Delta\gamma} = \frac{\text{known } \mu \text{ interval}}{\Delta\mu}$$

$$\frac{0.86 - \gamma}{0.86 - 0.914} = \frac{0.05 - 0.042}{0.05 - 0.01}$$

$$\Rightarrow \gamma = 0.87_1$$

12-13. (a) $\log \gamma = \dfrac{-0.51 \cdot 2^2 \cdot \sqrt{0.083}}{1 + (600\sqrt{0.083}/305)} = -0.375 \Rightarrow \gamma = 10^{-0.375} = 0.42_2$

(b) $\gamma = \left(\dfrac{0.083 - 0.05}{0.1 - 0.05}\right)(0.405 - 0.485) + 0.485 = 0.43_2$

12-14. (a) $\mu = 0.060$ M from KNO_3 (assuming that AgSCN has negligible solubility)

$[Ag^+]\gamma_{Ag^+}[SCN^-]\gamma_{SCN^-} = K_{sp}$

$[x](0.79)[x](0.80) = 1.1 \times 10^{-12} \Rightarrow x = [Ag^+] = 1.3 \times 10^{-6}$ M

The answer $[Ag^+] = 1.3$ μM confirms that AgSCN contributes a negligible amount to the ionic strength.

(b) $\mu = 0.060$ M from KSCN

$[Ag^+]\gamma_{Ag^+}[SCN^-]\gamma_{SCN^-} = K_{sp}$

$[x](0.79)\underbrace{[x + 0.060]}_{\approx 0.060}(0.80) = 1.1 \times 10^{-12} \Rightarrow x = [Ag^+] = 2.9 \times 10^{-11}$ M

The answer $x = 29$ pM confirms that x can be neglected in comparison with 0.060 0 M.

12-15. Ionic strength = 0.010 M (from HCl) + 0.040 M (from $KClO_4$ that gives $K^+ + ClO_4^-$) = 0.050 M. Using Table 12-1, $\gamma_{H^+} = 0.86$.

$$\text{pH} = -\log([H^+]\gamma_{H^+}) = -\log((0.010)(0.86)) = 2.07$$

12-16. At an ionic strength of 0.050 M, $\gamma_{H^+} = 0.86$ and $\gamma_{OH^-} = 0.81$.

$[H^+]\gamma_{H^+}[OH^-]\gamma_{OH^-} = (x)(0.86)(x)(0.81) = 1.0 \times 10^{-14}$

$\Rightarrow x = [H^+] = 1.2 \times 10^{-7}$ M

$pH = -\log([H^+]\gamma_{H^+}) = -\log((1.2 \times 10^{-7})(0.86)) = 6.99$

12-17. Ionic strength $= 0.010$ M from NaOH $+ 0.012$ M from $LiNO_3 = 0.022$ M

To find the activity coefficient of OH^-, we interpolate in Table 12-1:

$\mu = 0.01 \quad\quad 0.022 \quad\quad 0.05$
$OH^-: \gamma = 0.900 \quad ? \quad\quad 0.81$

$$\frac{\text{unknown } \gamma \text{ interval}}{\Delta\gamma} = \frac{\text{known } \mu \text{ interval}}{\Delta\mu}$$

$$\frac{0.81 - \gamma}{0.81 - 0.900} = \frac{0.05 - 0.022}{0.05 - 0.01}$$

$\Rightarrow \gamma_{OH^-} = 0.873$

$$\mathcal{A}_{H^+} = [H^+]\gamma_{H^+} = \frac{K_w}{[OH^-]\gamma_{OH^-}} = \frac{1.0 \times 10^{-14}}{(0.010)(0.873)} = 1.15 \times 10^{-12}$$

$pH = -\log \mathcal{A}_{H^+} = -\log(1.15 \times 10^{-12}) = 11.94$

If we had neglected activities, $pH \approx -\log[H^+] = -\log\dfrac{K_w}{[OH^-]} = 12.00$

12-18. (a) $Mn(OH)_2(s) \rightleftharpoons Mn^{2+} + 2 OH^- \quad\quad K_{sp} = 1.6 \times 10^{-13}$
$\quad\quad\quad\quad\quad\quad\quad\quad\quad\quad\quad x \quad\quad 2x$

$[Mn^{2+}][OH^-]^2 = K_{sp}$

$[x][2x]^2 = 1.6 \times 10^{-13} \Rightarrow x = [Mn^{2+}] = \left(\dfrac{1.6 \times 10^{-13}}{4}\right)^{1/3} = 3.4_2 \times 10^{-5}$ M

$[OH^-] = 2x = 6.8 \times 10^{-5}$ M

$[H^+] = K_w/[OH^-] = 1.46 \times 10^{-10}$ M $\Rightarrow pH = -\log[H^+] = 9.84$

(b) $Mn(OH)_2(s) \rightleftharpoons Mn^{2+} + 2 OH^- \quad\quad K_{sp} = 1.6 \times 10^{-13}$
$\quad\quad\quad\quad\quad\quad\quad\quad\quad\quad\quad x \quad\quad 2x$

Assuming $Mn(OH)_2$ gives negligible ions, $\mu = 0.075$ from $NaClO_4$.

$[Mn^{2+}]\gamma_{Mn^{2+}}[OH^-]^2\gamma_{OH^-}^2 = K_{sp}$

$[x](0.445)[2x]^2(0.785)^2 = 1.6 \times 10^{-13}$

$\Rightarrow x = [Mn^{2+}] = \left(\dfrac{1.6 \times 10^{-13}}{4(0.445)(0.785)^2}\right)^{1/3} = 5.2_6 \times 10^{-5}$ M

$$[\text{OH}^-] = 2x = 1.05_3 \times 10^{-4} \text{ M}$$

$$\mathcal{A}_{\text{OH}^-} = [\text{OH}^-]\gamma_{\text{OH}^-} = (1.0_{53} \times 10^{-4})(0.785) = 8.2_6 \times 10^{-5}$$

$$\text{pH} = -\log \mathcal{A}_{\text{H}^+} = -\log(K_w/\mathcal{A}_{\text{OH}^-}) = -\log(K_w/8.2_6 \times 10^{-5}) = 9.92$$

12-19. The ionic strength is 0.100 M from $(\text{CH}_3)_4\text{N}^+\text{IO}_3^-$, assuming that the concentration of ions derived from dissolution of $\text{Ba}(\text{IO}_3)_2$ is negligible.

$$\text{Ba}(\text{IO}_3)_2(s) \rightleftharpoons \text{Ba}^{2+} + 2\text{IO}_3^- \qquad K_{sp} = 1.5 \times 10^{-9}$$
$$\phantom{\text{Ba}(\text{IO}_3)_2(s) \rightleftharpoons} x \quad\;\; 2x + 0.100$$
$$\phantom{\text{Ba}(\text{IO}_3)_2(s) \rightleftharpoons x \quad\;\;} \approx 0.100$$

$$K_{sp} = 1.5 \times 10^{-9} = [\text{Ba}^{2+}]\gamma_{\text{Ba}^{2+}}[\text{IO}_3^-]^2 \gamma_{\text{IO}_3^-}^2$$
$$= x(0.38)(2x + 0.100)^2(0.775)^2 \approx x(0.38)(0.100)^2(0.775)^2$$
$$\Rightarrow x = [\text{Ba}^{2+}] = 6.6 \times 10^{-7} \text{ M}$$

Our assumption is verified:
$$[\text{IO}_3^-] \text{ from Ba}(\text{IO}_3)_2 = 2x = 1.32 \times 10^{-6} \text{ M} \ll 0.100 \text{ M from } (\text{CH}_3)_4\text{N}^+\text{IO}_3^-.$$

12-20. First, assume $\mu = 0$.
$$[\text{Pb}^{2+}]\gamma_{\text{Pb}^{2+}}[\text{F}^-]^2\gamma_{\text{F}^-}^2 = K_{sp} = 3.6 \times 10^{-8}$$
$$(x)\;(1)\;(2x)^2\;(1)^2 = 3.6 \times 10^{-8} \Rightarrow x = [\text{Pb}^{2+}] = 2.0_8 \times 10^{-3} \text{ M}$$

With this value of x, the ionic strength is
$$\mu = \tfrac{1}{2}\{[\text{Pb}^{2+}]\cdot(+2)^2 + [\text{F}^-]\cdot(-1)^2\}$$
$$= \tfrac{1}{2}\{[(2.08 \times 10^{-3})(2)^2 + (4.16 \times 10^{-3})(-1)^2\} = 6.24 \times 10^{-3} \text{ M}$$

Interpolation in Table 12-1 gives $\gamma_{\text{Pb}^{2+}} = 0.723$ and $\gamma_{\text{F}^-} = 0.920$.

Now we repeat the cycle:
$$(x)(0.723)(2x)^2(0.920)^2 = K_{sp} \Rightarrow x = [\text{Pb}^{2+}] = 2.45 \times 10^{-3} \text{ M}$$
$$\Rightarrow \mu = 7.35 \times 10^{-3} \text{ M} \Rightarrow \gamma_{\text{Pb}^{2+}} = 0.706 \text{ and } \gamma_{\text{F}^-} = 0.914$$

Another cycle gives
$$x(0.706)(2x)^2(0.914)^2 = K_{sp} \Rightarrow x = [\text{Pb}^{2+}] = 2.48 \times 10^{-3} \text{ M}$$
$$\Rightarrow \mu = 7.44 \times 10^{-3} \text{ M} \Rightarrow \gamma_{\text{Pb}^{2+}} = 0.704 \text{ and } \gamma_{\text{F}^-} = 0.913$$

Finally a self-consistent answer is found:
$$(x)(0.704)(2x)^2(0.913)^2 = K_{sp} \Rightarrow x = [\text{Pb}^{2+}] = 2.48 \times 10^{-3} \text{ M}$$

Answer: $[\text{Pb}^{2+}] = 2.5 \times 10^{-3} \text{ M}$

12-21. (a) We don't know the ionic strength, so we need to find it by iteration. We begin by assuming that the ionic strength is so low that the activity coefficients are 1. From the computed values of $[Ca^{2+}]$ and $[SO_4^{2-}]$, we will find the ionic strength and the activity coefficients. Then we use these activity coefficients to compute new values of $[Ca^{2+}]$ and $[SO_4^{2-}]$.

1. $[Ca^{2+}][SO_4^{2-}] = x^2 = K_{sp} = 2.4 \times 10^{-5} \Rightarrow x = 4.90 \times 10^{-3}$ M
$\Rightarrow \mu = 0.0196$ M $\Rightarrow \gamma_{Ca^{2+}} = 0.629$ and $\gamma_{SO_4^{2-}} = 0.608$

2. $[Ca^{2+}] \gamma_{Ca^{2+}} [SO_4^{2-}] \gamma_{SO_4^{2-}} = 2.4 \times 10^{-5}$
$(x)(0.629)(x)(0.608) = 2.4 \times 10^{-5} \Rightarrow x = 7.92 \times 10^{-3}$ M
$\Rightarrow \mu = 0.0317$ M $\Rightarrow \gamma_{Ca^{2+}} = 0.572$ and $\gamma_{SO_4^{2-}} = 0.543$

3. $(x)(0.572)(x)(0.543) = 2.4 \times 10^{-5} \Rightarrow x = 8.79 \times 10^{-3}$ M
$\Rightarrow \mu = 0.0352$ M $\Rightarrow \gamma_{Ca^{2+}} = 0.555$ and $\gamma_{SO_4^{2-}} = 0.525$

4. $(x)(0.555)(x)(0.525) = 2.4 \times 10^{-5} \Rightarrow x = 9.08 \times 10^{-3}$ M
$\Rightarrow \mu = 0.0363$ M $\Rightarrow \gamma_{Ca^{2+}} = 0.550$ and $\gamma_{SO_4^{2-}} = 0.518$

5. $(x)(0.550)(x)(0.518) = 2.4 \times 10^{-5} \Rightarrow x = 9.18 \times 10^{-3}$ M

Note: When I tried this same problem on a spreadsheet using the extended Debye-Hückel equation to compute activity coefficients (instead of the interpolation used above), the answer came out to 9.9 mM.

(b) The concentration of free Ca^{2+} is 9.9 mM. The rest of the dissolved material is probably the *ion pair*, $CaSO_4(aq)$.

12-22. $[H^+] + 2[Ca^{2+}] + [Ca(HCO_3)^+] + [Ca(OH)^+] + [K^+]$
$= [OH^-] + [HCO_3^-] + 2[CO_3^{2-}] + [ClO_4^-]$

12-23. $[H^+] = [OH^-] + [HSO_4^-] + 2[SO_4^{2-}]$

12-24. $[H^+] = [OH^-] + [H_2AsO_4^-] + 2[HAsO_4^{2-}] + 3[AsO_4^{3-}]$

$$H-O-\overset{\overset{O}{\|}}{\underset{\underset{O^-}{|}}{As}}-O^-$$

12-25. (a) $2[Mg^{2+}] + [H^+] = [Br^-] + [OH^-]$

(b) $2[Mg^{2+}] + [H^+] + [MgBr^+] = [Br^-] + [OH^-]$

12-26. $[CH_3CO_2^-] + [CH_3CO_2H] = 0.1$ M

12-27. (a) $0.20 \text{ M} = [Mg^{2+}]$ (c) $0.20 \text{ M} = [Mg^{2+}] + [MgBr^+]$

(b) $0.40 \text{ M} = [Br^-]$ (d) $0.40 \text{ M} = [Br^-] + [MgBr^+]$

12-28. (a) $[F^-] + [HF] = 2[Ca^{2+}]$

(b) $\underbrace{[F^-] + [HF] + 2[HF_2^-]}_{\text{Mol F}} = 2[Ca^{2+}]$

(1 mol of HF_2^- contains 2 mol F)

12-29. (a) There is 3/2 as much calcium as phosphorus in the solution. So,

$$2[Ca^{2+}] = 3\{[PO_4^{3-}] + [HPO_4^{2-}] + [H_2PO_4^-] + [H_3PO_4]\}$$

(b) There is 3/2 as much sulfur as iron in the solution. So, $3\{[Fe^{3+}] + [Fe(OH)^{2+}] + [Fe(OH)_2^+] + [FeSO_4^+]\} = 2\{[SO_4^{2-}] + [HSO_4^-] + [FeSO_4^+]\}$

12-30. $Y_{total} = \frac{3}{2} X_{total}$

$\underbrace{2[X_2Y_2^{2+}] + [X_2Y^{4+}] + 3[X_2Y_3] + [Y^{2-}]}_{Y_{total}} = \frac{3}{2}\underbrace{\{2[X_2Y_2^{2+}] + 2[X_2Y^{4+}] + 2[X_2Y_3]\}}_{X_{total}}$

Canceling like terms on both sides allows us to simplify the mass balance to $[Y^{2-}] = [X_2Y_2^{2+}] + 2[X_2Y^{4+}]$

12-31. Charge balance: $2[Mg^{2+}] + [H^+] = [OH^-]$ (1)

Mass balance: $[Mg^{2+}] = 4.0 \times 10^{-8} \text{ M}$ (2)

Equilibrium: $K_w = [H^+][OH^-]$ (3)

For another mass balance, we cannot write $[OH^-] = 2[Mg^{2+}]$ because OH^- comes from both $Mg(OH)_2$ and H_2O ionization. Setting $[H^+] = x$ and $[Mg^{2+}] = 4.0 \times 10^{-8}$ M in Equation 1 gives $[OH^-] = x + 8.0 \times 10^{-8}$. Putting this value into Equation 3 gives $K_w = (x)(x + 8.0 \times 10^{-8}) \Rightarrow x = [H^+] = 6.8 \times 10^{-8}$ M. Then we can say $[OH^-] = x + 8.0 \times 10^{-8} = 1.4_8 \times 10^{-7}$ M.

12-32. Charge balance: Invalid because pH is fixed.
Mass balance: $[R_3NH^+] + [R_3N] = [Br^-]$ (1)
Equilibria: $K_{sp} = [R_3NH^+][Br^-]$ (2)

$$K_a = \frac{[R_3N][H^+]}{[R_3NH^+]} \quad (3)$$

$K_w = [H^+][OH^-]$ (4)

If $[H^+] = 10^{-9.50}$, we can use Equation 3 to write

$$[R_3N] = \frac{K_a}{[H^+]}[R_3NH^+] = 7.27[R_3NH^+]$$

Putting this value into Equation 1 gives
$[R_3NH^+] + 7.27[R_3NH^+] = [Br^-]$
$8.27[R_3NH^+] = [Br^-]$

Putting this relation into Equation 2 gives

$$K_{sp} = [R_3NH^+][Br^-] = \left(\frac{[Br^-]}{8.27}\right)[Br^-] \Rightarrow [Br^-] = 5.8 \times 10^{-4} \text{ M}$$

This concentration must be equal to the solubility of $R_3NH^+Br^-$, because all Br^- originates from this salt.

12-33. (a) Reactions that would affect the solubility equilibrium $CaSO_4(s) \rightleftharpoons Ca^{2+} + SO_4^{2-}$ must consume Ca^{2+} or SO_4^{2-} to increase the amount of $CaSO_4$ that dissolves. Sulfate is not substantially protonated until the pH is lowered below 3. Ca^{2+} does not react to make much $CaOH^+$ until the pH is raised to near 12. Because no reactions consume Ca^{2+} or SO_4^{2-} in the pH range 3 to 8.5, the solubility of $CaSO_4$ is constant over this pH range.

(b) If the pH is lowered below 3, SO_4^{2-} becomes protonated to make HSO_4^-. If the pH is raised to 12, Ca^{2+} reacts to give $CaOH^+$. Both of these reactions consume products of the solubility equilibrium $CaSO_4(s) \rightleftharpoons Ca^{2+} + SO_4^{2-}$ and therefore increase the solubility of $CaSO_4$.

12-34. (a) Charge balance: Invalid because pH is fixed.
Mass balance: $[Ag^+] = [CN^-] + [HCN]$ (1)

Equilibria: $K_b = \dfrac{[HCN][OH^-]}{[CN^-]}$ (2)

$K_{sp} = [Ag^+][CN^-]$ (3)

$K_w = [H^+][OH^-]$ (4)

Because $[H^+] = 10^{-9.00}$, $[OH^-] = 10^{-5.00}$ M. Putting this value of $[OH^-]$ into Equation 2 gives

$$[HCN] = \frac{K_b[CN^-]}{[OH^-]} = \frac{(1.6 \times 10^{-5})[CN^-]}{1.0 \times 10^{-5}} = 1.6[CN^-]$$

Substituting into Equation 1 gives

$$[Ag^+] = [CN^-] + 1.6[CN^-] = 2.6[CN^-]$$

Substituting into Equation 3 gives $[Ag^+]\left(\frac{[Ag^+]}{2.6}\right) = K_{sp}$

$$\Rightarrow [Ag^+] = 2.4 \times 10^{-8} \text{ M}$$
$$[CN^-] = [Ag^+]/2.6 = 9.2 \times 10^{-9} \text{ M}$$
$$[HCN] = 1.6[CN^-] = 1.5 \times 10^{-8} \text{ M}$$

(b) With activities:

$$[Ag^+] = [CN^-] + [HCN] \qquad (1')$$

$$K_b = \frac{[HCN]\gamma_{HCN}[OH^-]\gamma_{OH^-}}{[CN^-]\gamma_{CN^-}} \qquad (2')$$

$$K_{sp} = [Ag^+]\gamma_{Ag^+}[CN^-]\gamma_{CN^-} \qquad (3')$$

$$K_w = [H^+]\gamma_{H^+}[OH^-]\gamma_{OH^-} \qquad (4')$$

Because pH = 9.00, $[OH^-]\gamma_{OH^-} = K_w/[H^+]\gamma_{H^+} = 10^{-5.00}$

Putting this value into Equation 2' gives

$$[HCN] = \frac{K_b \gamma_{CN^-}[CN^-]}{\gamma_{HCN}[OH^-]\gamma_{OH^-}}$$

$$= \frac{(1.6 \times 10^{-5})(0.755)[CN^-]}{1 \cdot 1.0 \times 10^{-5}} = 1.208[CN^-]$$

Here we have assumed $\gamma_{HCN} = 1$. Using the above relation in the mass balance (Equation 1') gives

$$[Ag^+] = 2.208[CN^-]$$

Substituting into Equation 3' gives $K_{sp} = [Ag^+](0.75)\left(\frac{[Ag^+]}{2.208}\right)(0.755)$

$$\Rightarrow [Ag^+] = 2.9 \times 10^{-8} \text{ M}$$
$$[CN^-] = \frac{[Ag^+]}{2.208} = 1.3 \times 10^{-8} \text{ M}$$
$$[HCN] = 1.208[CN^-] = 1.6 \times 10^{-8} \text{ M}$$

12-35. (a) Charge balance: Invalid because pH is fixed.

Mass balance: We are tempted to write $[OH^-] = 2[Pb^{2+}]$, but this expression is not true. OH^- also comes from water ionization and from the buffer used to fix the pH. We do not have enough information to write a mass balance.

Equilibria: $K = [Pb^{2+}][OH^-]^2 = 5.0 \times 10^{-16}$ (1)

$K_w = [H^+][OH^-] = 1.0 \times 10^{-14}$ (2)

If pH = 10.50, $[OH^-] = 10^{-3.50}$ M. Putting this value into Equation 1 gives $[Pb^{2+}] = K/(10^{-3.50})^2 = 5.0 \times 10^{-9}$ M

(b) Equilibria: $K = [Pb^{2+}][OH^-]^2 = 5.0 \times 10^{-16}$ (1')

$K_a = \dfrac{[PbOH^+][H^+]}{[Pb^{2+}]} = 2.5 \times 10^{-8}$ (2')

As in part (a), we find $[Pb^{2+}] = 5.0 \times 10^{-9}$ M with Equation (1'). Putting this value into Equation (2') gives

$[PbOH^+] = \dfrac{K_a[Pb^{2+}]}{[H^+]} = 4.0 \times 10^{-6}$ M

$[Pb]_{total} = [Pb^{2+}] + [PbOH^+] = 4.0 \times 10^{-6}$ M

(c) $K = [Pb^{2+}]\gamma_{Pb^{2+}}[OH^-]^2\gamma_{OH^-}^2 = [Pb^{2+}](0.455)\underbrace{(10^{-3.50})^2}_{[OH^-]\gamma_{OH^-} = 10^{-3.50}}$

$\Rightarrow [Pb^{2+}] = 1.1 \times 10^{-8}$ M

12-36. (a) $\alpha_{HA} = \dfrac{[H^+]}{[H^+] + K_a} = \dfrac{10^{-2.00}}{10^{-2.00} + 10^{-3.67}} = 0.98$

$\alpha_{A^-} = \dfrac{K_a}{[H^+] + K_a} = 0.021$

(b), (c)

	α_{HA}	α_{A^-}
pH = 3.00:	0.82	0.18
pH = 3.50:	0.60	0.40

12-37. (a) $\alpha_B = \dfrac{[B^+]}{[B] + [BH^+]} = \alpha_{A^-} = \dfrac{K_a}{[H^+] + K_a}$

$= \dfrac{6.3 \times 10^{-6}}{10^{-4.00} + 6.3 \times 10^{-6}} = 0.059$

$$\alpha_{BH^+} = \frac{[BH^+]}{[B] + [BH^+]} = \alpha_{HA} = \frac{[H^+]}{[H^+] + K_a}$$

$$= \frac{10^{-4.00}}{10^{-4.00} + 6.3 \times 10^{-6}} = 0.94$$

(b), (c)

	α_B	α_{BH^+}
pH = 5.00:	0.39	0.61
pH = 6.00:	0.86	0.14

12-38. (a) $K_a = K_w/K_b = 10^{-10.00}$

$$\alpha_B = \frac{K_a}{[H^+] + K_a} = \frac{10^{-10.00}}{10^{-9.00} + 10^{-10.00}} = 0.090_9$$

$$\alpha_{BH^+} = \frac{[H^+]}{[H^+] + K_a} = \frac{10^{-9.00}}{10^{-9.00} + 10^{-10.00}} = 0.90_9$$

(b), (c)

	α_B	α_{BH^+}
pH = 10.00:	0.50_0	0.50_0
pH = 10.30:	0.66_6	0.33_4

12-39.

	A	B	C	D	E
1	Fractional composition of monoprotic acid				
2					
3	Constants:	pH	[H+]	[HA]	[A-]
4	Ka =	2	1.00E-02	2.00E-01	2.10E-09
5	1.05E-10	2.5	3.16E-03	2.00E-01	6.64E-09
6	F =	3	1.00E-03	2.00E-01	2.10E-08
7	0.2	3.5	3.16E-04	2.00E-01	6.64E-08
8		4	1.00E-04	2.00E-01	2.10E-07
9		...	...	...	...
10		11.5	3.16E-12	5.85E-03	1.94E-01
11		12	1.00E-12	1.89E-03	1.98E-01
12	C4 = 10^-B4				
13	D4 = $A$7*C4/(C4+$A$5)				
14	E4 = $A$7*$A$5/(C4+$A$5)				

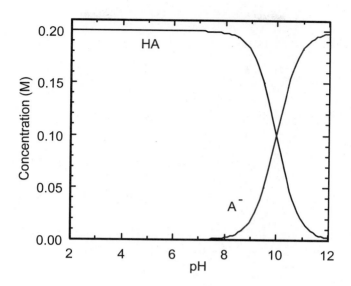

12-40.

	A	B	C	D	E	F	G
1	Fractional composition of diprotic acid						
2							
3	pK1 =	pH	[H+]	Denominator	Alpha(H2A)	Alpha(HA-)	Alpha(A2-)
4	3.02	1	1.00E-01	1.01E-02	9.91E-01	9.46E-03	3.13E-06
5	pK2 =	2	1.00E-02	1.10E-04	9.13E-01	8.71E-02	2.89E-04
6	4.48	3	1.00E-03	1.99E-06	5.03E-01	4.81E-01	1.59E-02
7	K1 =	4	1.00E-04	1.37E-07	7.29E-02	6.96E-01	2.31E-01
8	9.55E-04	5	1.00E-05	4.13E-08	2.42E-03	2.31E-01	7.66E-01
9	K2 =	6	1.00E-06	3.26E-08	3.07E-05	2.93E-02	9.71E-01
10	3.31E-05	7	1.00E-07	3.17E-08	3.15E-07	3.01E-03	9.97E-01
11		8	1.00E-08	3.16E-08	3.16E-09	3.02E-04	1.00E+00
12							
13	A8 = 10^-A4		D4= C4^2+$A$8*C4+$A$8*$A$10			F4 = $A$8*C4/D4	
14	C4 = 10^-B4		E4 = C4^2/D4			G4 = $A$8*$A$10/D4	

12-41. (a) $Pb^{2+} + H_2O \rightleftharpoons PbOH^+ + H^+$. Lead(II) ion behaves as a weak acid in water. If PbI_2 is dissolved in very pure water (without dissolved CO_2) and the pH is measured, the concentration of H^+ measured by the pH meter is equal to the concentration of $PbOH^+$.

(b) If the only equilibrium were $PbI_2(s) \rightleftharpoons Pb^{2+} + 2I^-$, then as extra I^- is added to solution (as KI, for example), the concentration of dissolved lead should decrease. If the reaction $Pb^{2+} + I^- \rightleftharpoons PbI^+$ occurs, then eventually adding more I^- would increase the dissolved lead. To provide evidence for the formation of PbI^+ and to measure its equilibrium constant for formation, we could measure the total concentration of dissolved lead (which would be $[Pb^{2+}] + [PbI^+]$) as a function of added KI.

12-42. The equilibrium constant is

$$HA \rightleftharpoons H^+ + A^- \qquad K_a = \frac{[H^+]\gamma_{H^+}[A^-]\gamma_{A^-}}{[HA]\gamma_{HA}}$$

From the table of activity coefficients, $\gamma_{H^+} = 0.83$ and $\gamma_{A^-} = 0.80$ at $\mu = 0.1$ M.

For HA, we estimate

$$\log \gamma_{HA} = k\mu = (0.2)(0.1) = 0.02, \text{ or } \gamma_{HA} = 10^{0.02} = 1.05$$

Putting these activity coefficients into the equilibrium expression gives

$$K_a \ (\mu = 0.1 \text{ M}) = \frac{[H^+]\gamma_{H^+}[A^-]\gamma_{A^-}}{[HA]\gamma_{HA}} = \frac{[H^+](0.83)[A^-](0.80)}{[HA](1.05)} = 0.63 \frac{[H^+][A^-]}{[HA]}$$

$$K_a \ (\mu = 0 \text{ M}) = \frac{[H^+]\gamma_{H^+}[A^-]\gamma_{A^-}}{[HA]\gamma_{HA}} = \frac{[H^+](1)[A^-](1)}{[HA](1)} = \frac{[H^+][A^-]}{[HA]}$$

Therefore,

$$\frac{[H^+][A^-]/[HA] \ (\text{at } \mu = 0.1 \text{ M})}{[H^+][A^-]/[HA] \ (\text{at } \mu = 0 \text{ M})} = 0.63$$

in agreement with the observed value of 0.63 ± 0.03.

CHAPTER 13
EDTA TITRATIONS

13-1. (a) (50.0 mL)(0.0100 mmol/mL) = 0.500 mmol Ca^{2+}, which requires 0.500 mmol EDTA = 10.0 mL EDTA.

(b) 0.500 mmol Al^{3+} requires the same amount of EDTA, 10.0 mL.

13-2. A back titration is necessary if the analyte precipitates in the absence of EDTA, if it reacts too slowly with EDTA, or if it blocks the indicator.

13-3. In a displacement titration, analyte displaces a metal ion from a complex. The displaced metal ion is then titrated. An example is the liberation of Ni^{2+} from $Ni(CN)_4^{2-}$ by the analyte Ag^+. The liberated Ni^{2+} is then titrated by EDTA to find out how much Ag^+ was present.

13-4. The Mg^{2+} in a solution of Mg^{2+} and Fe^{3+} can be titrated by EDTA if the Fe^{3+} is masked with CN^- to form $Fe(CN)_6^{3-}$, which does not react with EDTA.

13-5. Hardness refers to the total concentration of alkaline earth cations in water, which normally means $[Ca^{2+}] + [Mg^{2+}]$. Hardness gets its name from the reaction of these cations with soap to form insoluble curds. Temporary hardness, due to $Ca(HCO_3)_2$, is lost by precipitation of $CaCO_3(s)$ upon heating. Permanent hardness derived from other salts, such as $CaSO_4$, and is not affected by heat.

13-6. An auxiliary complexing agent forms a weak complex with analyte ion, thereby keeping it in solution without interfering with the EDTA titration. For example, NH_3 keeps Zn^{2+} in solution at high pH.

13-7. 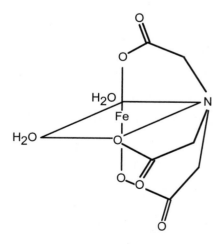 Nitrilotriacetic acid probably binds to Fe through three carboxyl oxygen atoms and one nitrogen atom. Because Fe^{3+} normally has six ligands, the other two sites are probably occupied by water (or OH^-, depending upon the pH).

13-8. The unknown was treated with 10.00 mL of 0.036 7 M EDTA, which contains (10.00 mL)(0.036 7 M) = 0.367 mmol of EDTA.

$$Fe^{3+} \;+\; EDTA \;\rightarrow\; Fe(EDTA) \;+\; EDTA$$
$$x \text{ mmol} \quad\quad 0.367 \text{ mmol} \quad\quad\quad\quad\quad\quad\quad 0.367 - x \text{ mmol}$$

Back titration required (2.37 mL)(0.046 1 M) = 0.109 mmol of Mg^{2+}.

$$Mg^{2+} \;+\; EDTA \;\rightarrow\; Mg(EDTA)$$
$$0.109 \text{ mmol} \quad 0.367 - x \text{ mmol} \quad 0.109 \text{ mmol}$$

The mol of Mg^{2+} required in the second reaction must equal the mol of excess EDTA from the first reaction

$$0.109 \text{ mmol } Mg^{2+} = 0.367 \text{ mmol EDTA} - x \text{ mmol } Fe^{3+}$$
$$x = 0.258 \text{ mmol } Fe^{3+}$$

The concentration of Fe^{3+} is 0.258 mmol/25.00 mL = 0.010 3 M. A quicker way to reach the same answer is to write

$$[Fe^{3+}] = \frac{\text{mmol EDTA} - \text{mmol } Mg^{2+}}{25.00 \text{ mL}} = \frac{0.367 - 0.109}{25.00 \text{ mL}} = 0.010\,3 \text{ M}$$

13-9.
Total EDTA = (25.0 mL)(0.045 2 M) = 1.130 mmol
− Mg^{2+} required = (12.4 mL)(0.012 3 M) = 0.153 mmol
Ni^{2+} + Zn^{2+} = 0.977 mmol

Zn^{2+} = EDTA displaced by 2,3-dimercapto-1-propanol
= (29.2 mL)(0.012 3 M) = 0.359 mmol

$\Rightarrow Ni^{2+} = 0.977 - 0.359 = 0.618$ mmol; $[Ni^{2+}] = \dfrac{0.618 \text{ mmol}}{50.0 \text{ mL}} = 0.012\,4$ M

$[Zn^{2+}] = \dfrac{0.359 \text{ mmol}}{50.0 \text{ mL}} = 0.007\,18$ M

13-10. (a) Unknown was treated with 10.00 mL of 0.040 0 M EDTA, which contains (10.00 mL)(0.040 0 M) = 0.400_0 mmol EDTA.

Excess EDTA after reaction with all metals required 5.05 mL of 0.026 2 M $ZnSO_4$ for back titration = (5.05 mL)(0.026 2 M) = 0.132_3 mmol Zn^{2+}.

Total mmol metal in unknown = mmol EDTA required for reaction with unknown = 0.400_0 mmol − 0.132_3 mmol = 0.267_7 mmol.

(b) The second titration required 5.81 mL of 0.026 2 M Zn^{2+} to titrate EDTA liberated from Hg^{2+}-EDTA complex = (5.81 mL)(0.026 2 M) = 0.152_2 mmol. There must have been 0.152_2 mmol Hg^{2+} (and 0.267_7 mmol − 0.152_2 mmol = 0.115_5 mmol of all other metal ions) in the unknown.

EDTA Titrations

(c) The table states that xylenol orange is yellow below pH 6.7 and metal complexes of xylenol orange are red regardless of pH. Before any Zn^{2+} is added, there is excess EDTA, so xylenol orange should be free and the color should be yellow. At the first end point, we expect to see red color. After adding KI, EDTA is liberated from Hg^{2+}, so the indicator should turn back to its yellow color because there is no metal-indicator complex present. At the second end point, the color should be red again.

13-11. The precipitation reaction is $Cu^{2+} + S^{2-} \rightarrow CuS(s)$.

Total Cu^{2+} used = (25.00 mL)(0.043 32 M) = 1.083 0 mmol
− Excess Cu^{2+} = (12.11 mL)(0.039 27 M) = 0.475 6 mmol
mmol of S^{2-} = 0.607 4 mmol

$[S^{2-}]$ = 0.607 4 mmol/25.00 mL = 0.024 30 M

13-12. For every mole of K^+ entering the first reaction, four moles of EDTA are produced in the second reaction.

moles of EDTA = moles of Zn^{2+} used in titration

$$[K^+] = \frac{\left(\frac{1}{4}\right)(\text{moles of } Zn^{2+})}{\text{volume of original sample}}$$

$$= \frac{\left(\frac{1}{4}\right)[28.73\,(\pm 0.03)\,\text{mL}][0.043\,7\,(\pm 0.000\,1)\,\text{M}]}{250.0\,(\pm 0.1)\,\text{mL}}$$

$$= \frac{[\frac{1}{4}(\pm 0\%)][28.73\,(\pm 0.104\%)][0.043\,7\,(\pm 0.229\%)]}{250.0\,(\pm 0.040\,0\%)}$$

$$= 1.256\,(\pm 0.255\%) \times 10^{-3}\,\text{M} = 1.256\,(\pm 0.003)\,\text{mM}$$

$$\%e = \sqrt{0^2 + (0.104)^2 + (0.229)^2 + (0.040\,0)^2}$$

13-13. Total $Fe^{3+} + Cu^{2+}$ in 25.00 mL = (16.06 mL)(0.050 83 M) = 0.816 3 mmol
Second titration:

millimoles EDTA used: (25.00 mL)(0.050 83 M) = 1.270 8
millimoles Pb^{2+} needed: (19.77 mL)(0.018 83 M) = 0.372 3

millimoles Fe^{3+} present: (difference) = 0.898 5

50.00 mL of unknown were used in the second titration, so the number of millimoles of Fe^{3+} in 25.00 mL is $\frac{1}{2}(0.898\,5) = 0.449\,2$. The millimoles of Cu^{2+} in 25.00 mL is 0.816 3 − 0.449 2 = 0.367 1 mmol/25.00 mL = 0.014 68 M.

13-14. Concentration of Ni^{2+} in standard solution is $\dfrac{(39.3 \text{ mL})(0.0130 \text{ M})}{30.0 \text{ mL}} = 0.0170 \text{ M}$.

The quantity of Ni^{2+} in 25.0 mL is $(25.0 \text{ mL})(0.0170 \text{ M}) = 0.425$ mmol.

Excess Ni = $(10.1 \text{ mL})(0.0130 \text{ M}) = 0.131$ mmol.

Therefore, $0.425 - 0.131 = 0.294$ mmol of Ni^{2+} reacted with CN^-.

Each mole of Ni^{2+} reacts with 4 mol CN^-, so there must have been

$4(0.294 \text{ mmol}) = 1.176$ mmol of CN^- in 12.7 mL.

$[CN^-] = 1.176$ mmol/12.7 mL = 0.0926 M

13-15.

	Total EDTA = (39.98 mL)(0.04500 M)	= 1.7991 mmol	
−	mmol Mg^{2+} = (10.26 mL)(0.02065 M)	= 0.2119 mmol	= 5.150 mg Mg
−	mmol Zn^{2+} = (15.47 mL)(0.02065 M)	= 0.3195 mmol	= 20.89 mg Zn
	mmol Mn^{2+}	= 1.2677 mmol	= 69.64 mg Mn

13-16. Total standard Ba^{2+} + Zn^{2+} added to the sulfate

$= (5.000 \text{ mL})(0.01463 \text{ M BaCl}_2) + (1.000 \text{ mL})(0.01000 \text{ M ZnCl}_2)$

$= 0.08315$ mmol

Total EDTA required = $(2.39 \text{ mL})(0.00963 \text{ M}) = 0.02302$ mmol

Original solid must have contained $0.08315 - 0.02302 = 0.06013$ mmol sulfur (which made 0.06013 mmol sulfate that precipitated 0.06013 mmol Ba^{2+}).

The mass of sulfur = $(0.06013 \text{ mmol})(32.065 \text{ mg/mmol}) = 1.92_8$ mg

wt% S = $100 \times (1.92_8 \text{ mg S}/5.89 \text{ mg sphalerite}) = 32.7$ wt%

Theoretical wt% S in pure ZnS = $100 \times (32.065 \text{ g S}/97.474 \text{ g ZnS}) = 32.90$ wt%

13-17. $\alpha_{Y^{4-}}$ is the fraction of all free EDTA in the form Y^{4-}.

(a) At pH 3.50:

$$\alpha_{Y^{4-}} = \frac{K_1 K_2 K_3 K_4 K_5 K_6}{\{[H^+]^6 + [H^+]^5 K_1 + [H^+]^4 K_1 K_2 + [H^+]^3 K_1 K_2 K_3 + [H^+]^2 K_1 K_2 K_3 K_4 + [H^+] K_1 K_2 K_3 K_4 K_5 + K_1 K_2 K_3 K_4 K_5 K_6\}}$$

Inserting the values $[H^+] = 10^{-3.50}$, $K_1 = 10^{-0.0}$, $K_2 = 10^{-1.5}$, $K_3 = 10^{-2.00}$, $K_4 = 10^{-2.69}$, $K_5 = 10^{-6.13}$, and $K_6 = 10^{-10.37}$ gives $\alpha_{Y^{4-}} = 2.7 \times 10^{-10}$.

(b) At pH 10.50, we find $\alpha_{Y^{4-}} = 0.57$

EDTA Titrations

13-18. (a) $Co^{2+} + 4NH_3 \rightleftharpoons Co(NH_3)_4^{2+}$ $\qquad \beta_4 = \dfrac{[Co(NH_3)_4^{2+}]}{[Co^{2+}][NH_3]^4}$

(b) $Co(NH_3)_3^{2+} + NH_3 \rightleftharpoons Co(NH_3)_4^{2+}$ $\qquad K_4 = \dfrac{[Co(NH_3)_4^{2+}]}{[Co(NH_3)_3^{2+}][NH_3]}$

From the relation $\beta_4 = K_1 K_2 K_3 K_4$, we can say $K_4 = \beta_4/K_1 K_2 K_3 = \beta_4/\beta_3 = 10^{5.07}/10^{4.43} = 10^{0.64}$ or $\log K_4 = 0.64$.

13-19. (a) $K_f' = \alpha_{Y^{4-}} K_f = 0.041 \times 10^{8.79} = 2.5 \times 10^7$

(b) $\begin{array}{cccc} Mg^{2+} & + & EDTA & \rightleftharpoons & MgY^{2-} \\ x & & x & & 0.050-x \end{array}$

$\dfrac{0.050-x}{x^2} = 2.5 \times 10^7 \Rightarrow [Mg^{2+}] = 4.5 \times 10^{-5}$ M

13-20. (a) mmol EDTA = mmol M^{n+}

$(V_e)(0.0500 \text{ M}) = (100.0 \text{ mL})(0.0500 \text{ M}) \Rightarrow V_e = 100.0$ mL

(b) Initial mmol $M^{n+} = (100.0 \text{ mL})(0.0500 \text{ M}) = 5.00$ mmol

At $\frac{1}{2}V_e$, there are 2.50 mmol M^{n+} in a total volume of 150.0 mL.

$[M^{n+}] = 2.50$ mmol/150.0 mL $= 0.0167$ M

(c) 0.041 (Table 13-3)

(d) $K_f' = (0.041)(10^{12.00}) = 4.1 \times 10^{10}$

(e) At the equivalence point, essentially all of the metal is converted to the EDTA complex: $[MY^{n-4}] = 5.00$ mmol/200.0 mL $= 0.0250$ M. A tiny amount of the complex, which we will call x, dissociates to M^{n+} and EDTA.

$\begin{array}{ccc} MY^{n-4} & \rightleftharpoons & M^{n+} + EDTA \\ 0.0250-x & & x \qquad x \end{array}$

$K_f' = \dfrac{[MY^{n-4}]}{[M^{n+}][EDTA]} = \dfrac{0.0250-x}{x^2}$

$= 4.1 \times 10^{10} \Rightarrow x = [M^{n+}] = 7.8 \times 10^{-7}$ M

(f) Excess EDTA = 10.0 mL

$[EDTA] = \dfrac{(0.0500 \text{ M})(10.0 \text{ mL})}{210.0 \text{ mL}} = 2.38 \times 10^{-3}$ M

$[MY^{n-4}] = \dfrac{5.00 \text{ mmol}}{210.0 \text{ mL}} = 2.38 \times 10^{-2}$ M

$\dfrac{[MY^{n-4}]}{[M^{n+}][EDTA]} = \dfrac{(2.38 \times 10^{-2})}{[M^{n+}](2.38 \times 10^{-3})} = 4.1 \times 10^{10} \Rightarrow [M^{n+}] = 2.4 \times 10^{-10}$ M

13-21. Titration reaction: $Mn^{2+} + EDTA \rightleftharpoons MnY^{2-}$ Equivalence volume = 50.0 mL

$$K'_f = \alpha_{Y^{4-}} K_f = (1.8 \times 10^{-5})(10^{13.89}) = 1.4 \times 10^9$$

Sample calculations:

20.0 mL: Total mmol Mn = (0.0200 M)(25.0 mL) = 0.500 mmol

Total mmol EDTA = (0.0100 M)(20.0 mL) = 0.200 mmol

mmol Mn^{2+} = 0.500 − 0.200 = 0.300 mmol

$[Mn^{2+}]$ = 0.300 mmol/45.0 mL = 6.67×10^{-3} M

$pMn^{2+} = -\log[Mn^{2+}] = 2.18$

50.0 mL: The formal concentration of MnY^{2-} is

$[MnY^{2-}]$ = 0.500 mmol/75.0 mL = 0.00667 M

$$\begin{array}{ccccc} Mn^{2+} & + & EDTA & \rightleftharpoons & MnY^{2-} \\ x & & x & & 0.00667 - x \end{array}$$

$$\frac{0.00667 - x}{x^2} = \alpha_{Y^{4-}} K_f \Rightarrow x = 2.1_8 \times 10^{-6} \Rightarrow pMn^{2+} = 5.66$$

60.0 mL: There are 10.0 mL of excess EDTA.

$$[EDTA] = \frac{(0.0100\ M)(10.0\ mL)}{85.0\ mL} = 1.17_6 \times 10^{-3}\ M$$

$[MnY^{2-}]$ = 0.500 mmol/85.0 mL = 5.88×10^{-3} M

$$[Mn^{2+}] = \frac{[MnY^{2-}]}{[EDTA]K'_f} = \frac{[5.88 \times 10^{-3}]}{[1.17_6 \times 10^{-3}](1.4 \times 10^9)} = 3.5_7 \times 10^{-9}$$

$$\Rightarrow pMn^{2+} = 8.45$$

Volume (mL)	pMn^{2+}	Volume	pMn^{2+}	Volume	pMn^{2+}
0	1.70	49.0	3.87	50.1	6.46
20.0	2.18	49.9	4.87	55.0	8.15
40.0	2.81	50.0	5.66	60.0	8.45

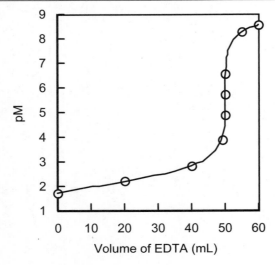

EDTA Titrations

13-22. Titration reaction: $Ca^{2+} + EDTA \rightleftharpoons CaY^{2-}$; equivalence volume = 50.0 mL.
$$K_f' = \alpha_{Y^{4-}} K_f = (0.30)(10^{10.65}) = 1.3_4 \times 10^{10}$$

Sample calculations:

20.0 mL: Total mmol EDTA = $(0.0200 \text{ M})(25.0 \text{ mL}) = 0.500$ mmol
mmol Ca^{2+} added = $(0.0100 \text{ M})(20.0 \text{ mL}) = 0.200$ mmol
mmol EDTA unreacted = $0.500 - 0.200 = 0.300$ mmol
$[EDTA] = 0.300 \text{ mmol}/45.0 \text{ mL} = 6.67 \times 10^{-3}$ M
$[CaY^{2-}] = 0.200 \text{ mmol}/45.0 \text{ mL} = 4.44 \times 10^{-3}$ M
$[Ca^{2+}] = \dfrac{[CaY^{2-}]}{[EDTA]K_f'} = 4.9_7 \times 10^{-11}$ M $\Rightarrow pCa^{2+} = 10.30$

50.0 mL: The formal concentration of CaY^{2-} is
$[CaY^{2-}] = 0.500 \text{ mmol}/75.0 \text{ mL} = 0.00667$ M

$$\begin{array}{ccccc} Ca^{2+} & + & EDTA & \rightleftharpoons & CaY^{2-} \\ x & & x & & 0.00667 - x \end{array}$$

$\dfrac{0.00667 - x}{x^2} = \alpha_{Y^{4-}} K_f \Rightarrow x = 7.0_5 \times 10^{-7}$ M $\Rightarrow pCa^{2+} = 6.15$

50.1 mL: There is an excess of 0.1 mL of Ca^{2+}.
$[Ca^{2+}] = \dfrac{(0.0100 \text{ M})(0.1 \text{ mL})}{75.1 \text{ mL}} = 1.33 \times 10^{-5}$ M $\Rightarrow pCa^{2+} = 4.88$

Volume (mL)	pCa^{2+}	Volume	pCa^{2+}	Volume	pCa^{2+}
0	(∞)	49.0	8.44	50.1	4.88
20.0	10.30	49.9	7.43	55.0	3.20
40.0	9.52	50.0	6.15	60.0	2.93

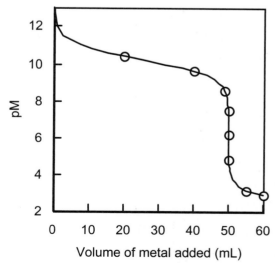

13-23. The addition of EDTA (Y) to metal (M) is analogous to the addition of a weak base (A^-) to a strong acid (H^+). EDTA behaves as the weak base and metal behaves as the strong acid.

$$M + Y \rightleftharpoons MY$$
$$H^+ + A^- \rightleftharpoons HA$$

Prior to the equivalence point, each increment of Y (A^-) is used up by reaction with excess M (H^+) and there is some M (H^+) left over. pM (pH) is determined by the concentration of excess M (H^+).

At the equivalence point, most of the reactants are converted to MY (HA), which slightly dissociates to give a small concentration of M (H^+).

Past the equivalence point, we are adding excess Y (A^-) to a solution of MY (HA). We have a a buffer whose concentration of M (H^+) is determined by the mixture of Y and MY (A^- and HA).

13-24. Titration reaction: $Cu^{2+} + EDTA \rightleftharpoons CuY^{2-}$
$$K'_f = \alpha_{Y^{4-}} K_f = (2.9 \times 10^{-7})(10^{18.78}) = 1.7_5 \times 10^{12}$$

The equivalence point is 50.0 mL.

Sample calculations:

20.0 mL: Total mmol Cu = (0.080$_0$ M)(25.0 mL) = 2.00$_0$ mmol

Total mmol EDTA = (0.040$_0$ M)(20.0 mL) = 0.800 mmol
mmol Cu^{2+} = 2.00$_0$ – 0.800 = 1.20$_0$ mmol
$[Cu^{2+}]$ = 1.20$_0$ mmol/45.0 mL = 2.67×10^{-2} M
$pCu^{2+} = -\log[Cu^{2+}] = 1.57$

50.0 mL: The formal concentration of CuY^{2-} is

$[CuY^{2-}]$ = 2.00 mmol/75.0 mL = 0.026 7 M

$$\begin{array}{ccc} Cu^{2+} + & EDTA \rightleftharpoons & CuY^{2-} \\ x & x & 0.026\,7 - x \end{array}$$

$$\frac{0.026\,7 - x}{x^2} = \alpha_{Y^{4-}} K_f \Rightarrow x = 1.2_3 \times 10^{-7} \Rightarrow pCu^{2+} = 6.91$$

51.0 mL: There is 1.0 mL of excess EDTA.

$$[EDTA] = \frac{(0.040\,0\text{ M})(1.0\text{ mL})}{76.0\text{ mL}} = 5.2_6 \times 10^{-4} \text{ M}$$

$[CuY^{2-}]$ = 2.00 mmol/76.0 mL = 2.63×10^{-2} M

$$[Cu^{2+}] = \frac{[CuY^{2-}]}{[EDTA]K'_f} = \frac{[2.63 \times 10^{-2}]}{[5.2_6 \times 10^{-4}](1.7_5 \times 10^{12})} = 2.8_6 \times 10^{-11} \text{ M}$$

$\Rightarrow pCu^{2+} = 10.54$

Volume (mL)	pCu^{2+}	Volume	pCu^{2+}	Volume	pCu^{2+}
0	1.10	49.0	3.27	51.0	10.54
20.0	1.57	50.0	6.91	55.0	11.24
40.0	2.21				

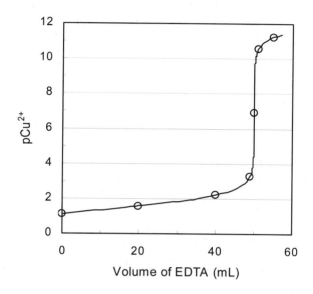

13-25. At pH 7.00, $\alpha_{Y^{4-}} = 3.8 \times 10^{-4}$ and
$$K'_f = \alpha_{Y^{4-}} K_f = (3.8 \times 10^{-4})(10^{18.78}) = 2.2_9 \times 10^{15}$$

The calculations prior to the equivalence point are identical at pH 5 and at pH 7, and the values of pCu^{2+} are identical. The pH only affects the calculations at and beyond the equivalence point.

50.0 mL: The formal concentration of CuY^{2-} is
$$[CuY^{2-}] = 2.00 \text{ mmol}/75.0 \text{ mL} = 0.026\,7 \text{ M}$$

$$\begin{array}{ccccc} Cu^{2+} & + & EDTA & \rightleftharpoons & CuY^{2-} \\ x & & x & & 0.026\,7 - x \end{array}$$

$$\frac{0.026\,7 - x}{x^2} = \alpha_{Y^{4-}} K_f \Rightarrow x = 3.4 \times 10^{-9} \Rightarrow pCu^{2+} = 8.47$$

51.0 mL: There is 1.0 mL of excess EDTA.
$$[EDTA] = \frac{(0.040\,0 \text{ M})(1.0 \text{ mL})}{76.0 \text{ mL}} = 5.2_6 \times 10^{-4} \text{ M}$$

$$[CuY^{2-}] = 2.00 \text{ mmol}/76.0 \text{ mL} = 2.63 \times 10^{-2} \text{ M}$$

$$[Cu^{2+}] = \frac{[CuY^{2-}]}{[EDTA]K'_f} = \frac{[2.63 \times 10^{-2}]}{[5.2_6 \times 10^{-4}](2.2_9 \times 10^{15})} = 2.1_8 \times 10^{-14}$$

$$\Rightarrow pCu^{2+} = 13.66$$

Volume (mL)	pCu^{2+}	Volume	pCu^{2+}	Volume	pCu^{2+}
0	1.10	49.0	3.27	51.0	13.66
20.0	1.57	50.0	8.47	55.0	14.36
40.0	2.21				

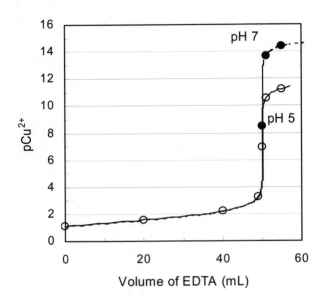

The fraction of EDTA in the form Y^{4-} increases from 2.9×10^{-7} at pH 5 to 3.8×10^{-4} at pH 7. This increase of $\alpha_{Y^{4-}}$ increases the extent of the reaction with EDTA at elevated pH and decreases the amount of free metal ion in solution beyond the equivalence point.

13-26. [Ca^{2+}] = $10^{-9.00}$ M, so essentially all calcium in solution is CaY^{2-}.

$$[CaY^{2-}] = \frac{1.95 \text{ g}}{(200.12 \text{ g/mol})(0.500 \text{ L})} = 0.019_9 \text{ M}$$

$$K_f' = \alpha_{Y^{4-}} K_f = (0.041)(10^{10.65}) = 1.8_3 \times 10^9$$

$$= \frac{[CaY^{2-}]}{[EDTA][Ca^{2+}]} = \frac{(0.019_9)}{[EDTA](10^{-9.00})} \Rightarrow [EDTA] = 0.010_{65} \text{ M}$$

Total EDTA needed = mol CaY^{2-} + mol free EDTA
  = (0.019$_9$ M)(0.500 L) + (0.010$_{65}$ M)(0.500 L) = 0.015$_{07}$ mol
  = 5.6$_1$ g Na$_2$H$_2$EDTA·2H$_2$O

13-27. (a)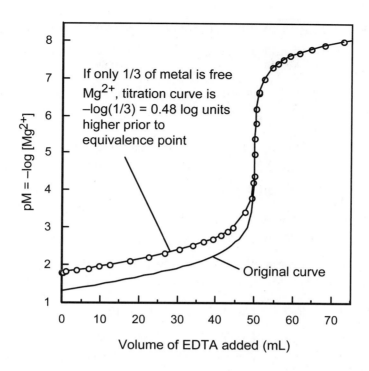

(b) There is no difference between the two curves past the equivalence point. We know that [MgY^{2-}] and [EDTA] are the same in both cases. From the effective formation constant,

$$K_f' = \frac{[\text{MgY}^{2-}]}{[\text{Mg}^{2+}][\text{EDTA}]}$$

we can be sure that if [MgY^{2-}] and [EDTA] are unchanged, [Mg^{2+}] must be the same. Even if 2/3 of the magnesium not bound to EDTA is tied up with NH$_3$ and OH$^-$, the concentration of [Mg^{2+}] must be the one value that satisfies the equilibrium constant, K_f'. In order to provide the extra Mg^{2+} to be bound to NH$_3$ and OH$^-$, a little bit of the MgY^{2-} dissociates.

13-28.

Neutral H$_5$DTPA has two carboxylic acid protons and three ammonium protons. We are not given the pK_a values, but, by analogy with EDTA, we expect carboxyl pK_a values to be below ~3 and ammonium pK_a values to be above ~6. At pH 14, we expect all acidic protons of DTPA to be dissociated, so the predominant species will be $DTPA^{5-}$. At pH 3-4, the nitrogen atoms should be protonated, but the carboxyl groups should be deprotonated. The predominant species is probably H_3DTPA^{2-}.

For HSO_4^-, $pK_a = 2.0$. At pH 14 and at pH 3, sulfate should mainly be in the form SO_4^{2-}.

At pH 14, $DTPA^{5-}$ is apparently a strong enough ligand to chelate Ba^{2+} and dissolve $BaSO_4(s)$. At pH 3–4, H_3DTPA^{2-} is not a strong enough ligand to dissolve $BaSO_4(s)$. An equivalent statement is that H^+ at a concentration of 10^{-3}–10^{-4} M competes with Ba^{2+} for binding sites on DTPA, but H^+ at a concentration of 10^{-14} M does not compete with Ba^{2+} for binding sites on DTPA.

Now that you have seen my reasoning, I'll provide some more information. The pK_a values for DTPA, beginning with the fully protonated H_8DTPA^{3+} are

H_8DTPA^{3+}	$pK_1 = -0.1$	CO_2H	H_4DTPA^-	$pK_5 = 2.7$	CO_2H
H_7DTPA^{2+}	$pK_2 = 0.7$	CO_2H	H_3DTPA^{2-}	$pK_6 = 4.3$	NH^+
H_6DTPA^+	$pK_3 = 1.6$	CO_2H	H_2DTPA^{3-}	$pK_7 = 8.6$	NH^+
H_5DTPA	$pK_4 = 2.0$	CO_2H	$HDTPA^{4-}$	$pK_8 = 10.5$	NH^+

As pH is lowered from 14, the three nitrogen atoms are 50% protonated at pH 10.5, 8.6, and 4.3. The third nitrogen atom is not quite fully protonated at pH 3–4. The predominant species is H_3DTPA^{2-}, as I guessed correctly. The species H_4DTPA^- and H_2DTPA^{3-} are also present to some extent in the pH range 3–4.

CHAPTER 14
ELECTRODE POTENTIALS

14-1. (a) Electric charge (coulombs) is a quantity of positive or negative particles. Current (amperes) is the quantity of charge per second moving past a point. Electric potential (volts) measures the work that can be done by (or must be done to) an electric charge as it moves from one point to another.

(b) $1/(1.602 \times 10^{-19}$ C/electron$) = 6.242 \times 10^{18}$ electrons/C

(c) $F \approx 96\,485$ C/mol

14-2. $\underset{\text{Oxidant Fe}(+3)}{Fe_2O_3}$ + $\underset{\text{Reductant Al}(0)}{2Al}$ → $\underset{\text{Fe}(0)}{2Fe}$ + $\underset{\text{Al}(+3)}{Al_2O_3}$

14-3. $\underset{\text{Reductant Na}(0)}{Na(s)}$ + $\underset{\substack{\text{Oxidant H}(+1)\\\text{O}(-2)}}{H_2O}$ ⇌ $\underset{\text{Na(I)}}{Na^+}$ + $\underset{\substack{\text{H}(+1)\\\text{O}(-2)}}{OH^-}$ + $\underset{\text{H}(0)}{\tfrac{1}{2}H_2(g)}$

Sodium is oxidized from Na(0) to Na(I). Half of the hydrogen from H_2O is reduced from H(I) to H(0). Oxygen is unchanged as O(–II) in H_2O and in OH^-.

$Na(s) \rightleftharpoons Na^+ + e^-$

$H_2O + e^- \rightleftharpoons OH^- + \tfrac{1}{2}H_2(g)$

14-4. (a) $\underset{\text{Oxidant Te}(+4)}{TeO_3^{2-}}$ + $3H_2O$ + $4e^-$ ⇌ $\underset{\text{Te}(0)}{Te(s)}$ + $6\,OH^-$

$\underset{\text{Reductant S}(+3)}{S_2O_4^{2-}}$ + $4\,OH^-$ ⇌ $\underset{\text{S}(+4)}{2SO_3^{2-}}$ + $2H_2O$ + $2e^-$

(b) (1.00 g Te)/(127.60 g/mol) = 7.84 mmol, which requires 4 × 7.84 = 31.3 mmol of electrons. $(3.13 \times 10^{-2}$ mol $e^-)(9.649 \times 10^4$ C/mol$)$
$= 3.02 \times 10^3$ C

(c) Current = $(3.02 \times 10^3$ C$)/(3\,600$ s$) = 0.840$ A

14-5. (a) $\underset{\text{Oxidant C}(+1)}{C_2HCl_3}$ + $3H^+$ + $6e^-$ ⇌ $\underset{\text{C}(-2)}{C_2H_4}$ + $3Cl^-$

$\underset{\text{Reductant Fe}(0)}{Fe}$ ⇌ $\underset{\text{Fe}(2+)}{Fe^{2+}}$ + $2e^-$

Net reaction: $3Fe + C_2HCl_3 + 3H^+ \rightarrow 3Fe^{2+} + C_2H_4 + 3Cl^-$
FM 131.39

129

(b) 17 kg C_2HCl_3/(131.39 g/mol) = 129.4 mol C_2HCl_3

Each mole of C_2HCl_3 requires 3 moles Fe: 3 × 129.4 mol = 388.2 mol Fe

(388.2 mol Fe)(55.845 g/mol) = 21.68 kg Fe

Fraction of Fe consumed in reaction = 21.68 kg/480 kg = 4.5%

(c) Rate of reaction = 129.4 mol C_2HCl_3/150 days. We need to convert this to a rate in mol/s and then to coulomb/s. The number of seconds in 150 days is (60 s/min)(60 min/h)(24 h/day)(150 day) = 1.296×10^7 s. The reaction rate is 129.4 mol/(1.296×10^7 s) = 9.98×10^{-6} mol/s. Each mole consumes 6 electrons, so the rate at which charge is transferred is

$$\left(9.98 \times 10^{-6} \frac{\text{mol } C_2HCl_3}{\text{s}}\right)\left(6 \frac{\text{mol e}^-}{\text{mol } C_2HCl_3}\right)\left(9.649 \times 10^4 \frac{C}{\text{mol e}^-}\right) = 5.78 \frac{C}{\text{s}} = 5.78 \text{ A}$$

14-6. $q = nF = (1.00 \times 10^{-6}$ mol$)(9.649 \times 10^4$ C/mol$) = 0.096\,49$ C

Work = Eq = (0.165 V)(0.096 49 C) = 0.0159 J

14-7. (a) I = coulombs/s. Every mole of O_2 accepts 4 moles of e^-.

(16 mol O_2/day)(4 e^-/mol O_2) = 64 mol e^-/day

(64 mol e^-/day)(1 day/8.64×10^4 s) = 7.41×10^{-4} mol e^-/s

(7.41×10^{-4} mol e^-/s)(9.649×10^4 C/mol e^-) = 71.5 C/s = 71.5 A

(b) Electric charge/day = (64 mol e^-)(9.649×10^4 C/mol e^-) = 6.18×10^6 C

Work = Eq = (1.1 V)(6.18×10^6 C) = 6.8×10^6 J

14-8. (a)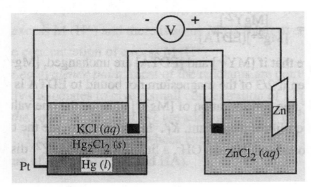

(b) reduction half-reactions:

$Hg_2Cl_2(s) + 2e^- \rightleftharpoons 2Hg(l) + 2Cl^-$

$Zn^{2+} + 2e^- \rightleftharpoons Zn(s)$

(c) Electrons flow from less positive to more positive potential, which is from Zn (–0.75 V) to Pt (+0.25 V) in this cell. Therefore, oxidation is taking

place at the Zn electrode, which must be the anode. Reduction occurs at Pt (really at Hg), which must be the cathode. The net cell reaction is
$$Hg_2Cl_2(s) + Zn(s) \rightleftharpoons 2Hg(l) + Zn^{2+} + 2Cl^-$$

14-9.

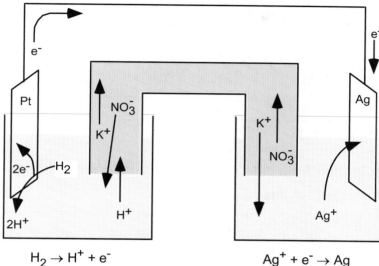

$H_2 \to H^+ + e^-$ $\qquad$ $Ag^+ + e^- \to Ag$

14-10. (a) right half-cell: $E_+ = \left(0.22 - \dfrac{0.059\,16}{2} \log[Cl^-]^2\right) = 0.281_2$ V

left half-cell: $E_- = \left(-0.350 - \dfrac{0.059\,16}{2} \log[F^-]^2\right) = -0.290_8$ V

$E = E_+ - E_- = 0.281_2 - (-0.290_8) = 0.572$ V

(b) $[Pb^{2+}] = K_{sp}$ (for PbF_2)/$[F^-]^2 = (3.6 \times 10^{-8})/(0.10)^2 = 3.6 \times 10^{-6}$ M
$[Ag^+] = K_{sp}$ (for $AgCl$)/$[Cl^-] = (1.8 \times 10^{-10})/(0.10) = 1.8 \times 10^{-9}$ M

right half-cell: $E_+ = \left(0.799 - \dfrac{0.059\,16}{2} \log\dfrac{1}{[Ag^+]^2}\right) = 0.281_2$ V

left half-cell: $E_- = \left(-0.126 - \dfrac{0.059\,16}{2} \log\dfrac{1}{[Pb^{2+}]}\right) = -0.287_0$ V

$E = E_+ - E_- = 0.281_2 - (-0.287_0) = 0.568$ V

The disagreement between the two results (0.572 and 0.568 V) is probably within the uncertainty of the standard potentials and solubility products.

14-11. (a) $Pt(s) \mid Cr^{2+}(aq), Cr^{3+}(aq) \parallel Tl^+(aq) \mid Tl(s)$

(b) right half-cell: $Tl^+ + e^- \rightleftharpoons Tl(s)$ $\qquad$ $E^\circ_+ = -0.336$ V
left half-cell: $Cr^{3+} + e^- \rightleftharpoons Cr^{2+}$ $\qquad$ $E^\circ_- = -0.42$ V

right half-cell: $E_+ = \left(-0.336 - 0.059\,16 \log\dfrac{1}{[Tl^+]}\right) = -0.336$ V left half-cell: $E_- = \left(-0.42 - 0.05916 \log\frac{[Cr^{2+}]}{[Cr^{3+}]}\right) = -0.42$ V

$E = E_+ - E_- = -0.336 - (-0.42) = 0.08_4$ V

(c)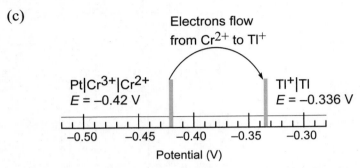

Electrons flow from Pt to Tl. Oxidation takes place at the anode, which is the Pt electrode in this cell.

(d) $Tl^+ + Cr^{2+} \rightleftharpoons Tl(s) + Cr^{3+}$

14-12. (a) right half-cell: $Hg_2Cl_2(s) + 2e^- \rightleftharpoons 2Hg(l) + 2Cl^-$ $E_+^\circ = 0.268$ V

left half-cell: $2H^+ + 2e^- \rightleftharpoons H_2(g)$ $E_-^\circ = 0$ V

right half-cell: $E_+ = \left(0.268 - \frac{0.05916}{2}\log[Cl^-]^2\right)$

$= \left(0.268 - \frac{0.05916}{2}\log[0.200]^2\right) = 0.309_4$ V

left half-cell: $E_- = \left(0 - \frac{0.05916}{2}\log\frac{P_{H_2}}{[H^+]^2}\right)$

$= \left(0 - \frac{0.05916}{2}\log\frac{0.100}{[10^{-2.54}]^2}\right) = -0.120_7$ V

(b)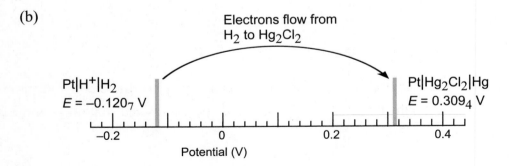

Electrons flow from the hydrogen electrode to the mercury electrode. Oxidation takes place at the hydrogen electrode, so the hydrogen electrode is the anode.

(c) $E = E_+ - E_- = 0.309_4 - (-0.120_7) = 0.430$ V

14-13. (a) right half-cell: $Al^{3+} + 3e^- \rightleftharpoons Al(s)$ $E°_+ = -1.677$ V
left half-cell: $Br_2(l) + 2e^- \rightleftharpoons 2Br^-$ $E°_- = 1.078$ V

right half-cell: $E_+ = \left(-1.677 - \dfrac{0.059\,16}{3} \log \dfrac{1}{[Al^{3+}]}\right)$

$= \left(-1.677 - \dfrac{0.059\,16}{3} \log \dfrac{1}{[0.010]}\right) = -1.716_4$ V

left half-cell: $E_- = \left(1.078 - \dfrac{0.059\,16}{2} \log [Br^-]^2\right)$

$= \left(1.078 - \dfrac{0.059\,16}{2} \log [0.10]^2\right) = 1.137_2$ V

$E = E_+ - E_- = -1.716_4 - 1.137_2 = -2.854$ V

(b)

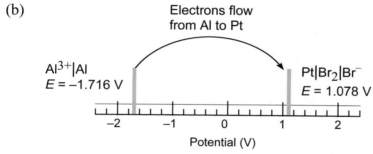

Electrons flow from Al to Pt. Reduction occurs at Pt and oxidation occurs at Al. The spontaneous reaction is $\tfrac{3}{2} Br_2(l) + Al(s) \rightleftharpoons 3Br^- + Al^{3+}$.

(c) 14.3 mL of Br_2 = 44.6 g = 0.279 mol of Br_2. 12.0 g of Al = 0.445 mol of Al. The reaction requires 3/2 mol Br_2 for every mole of Al, so 0.445 mol Al requires (3/2)(0.445 mol) = 0.668 mol Br_2. There are only 0.279 mol Br_2, so Br_2 will be used up first.

(d) (0.231 mL of Br_2)(3.12 g/mL) = 0.721 g of Br_2 = 4.51×10^{-3} mol Br_2
(4.51×10^{-3} mol Br_2)(2 mol e^-/mol Br_2) = 9.02×10^{-3} mol e^-
(9.02×10^{-3} mol e^-) (9.649×10^4 C/mol e^-) = 870 C
Work = $E \cdot q$ = (1.50 V)(870 C) = 1.31 kJ

(e) 2.89×10^{-4} A = 2.89×10^{-4} C/s
(2.89×10^{-4} C/s)/(9.649×10^4 C/mol e^-) = 2.99×10^{-9} mol e^-/s
Each mole of electrons comes from 1/3 mol Al:
(2.99×10^{-9} mol e^-/s)(1/3 mol Al/mol e^-) = 9.97×10^{-10} mol Al/s
(9.97×10^{-10} mol Al/s)(26.98 g/mol Al) = 2.69×10^{-8} g Al/s

14-14. The strategy is to find two half reactions whose difference is the desired net reaction. The reduction half reaction provides E_+° and the reaction that will be reversed gives E_-°.

(a) $\quad Cu^{2+} + e^- \rightleftharpoons Cu^+ \qquad\qquad E_+^\circ = 0.161$ V

$\quad\;\; -\;\; Cu^+ + e^- \rightleftharpoons Cu(s) \qquad\qquad E_-^\circ = 0.518$ V

$\overline{\quad\;\; Cu(s) + Cu^{2+} \rightleftharpoons 2Cu^+ \qquad\quad E^\circ = E_+^\circ - E_-^\circ = -0.357\text{ V}}$

$\quad K = 10^{1E^\circ/0.05916} = 9.2 \times 10^{-7}$

(b) $\quad 2F_2 + 4e^- \rightleftharpoons 4F^- \qquad\qquad E_+^\circ = 2.890$ V

$\quad\;\; -\;\; F_2O + 2H^+ + 4e^- \rightleftharpoons H_2O + 2F^- \qquad E_-^\circ = 2.168$ V

$\overline{\quad\;\; 2F_2 + H_2O \rightleftharpoons F_2O + 2F^- + 2H^+ \qquad E^\circ = 0.722\text{ V}}$

$\quad K = 10^{4E^\circ/0.05916} = 7 \times 10^{48}$

14-15. (a) $\quad \frac{1}{2}O_2(g) + 2H^+ + 2e^- \rightleftharpoons H_2O(l) \qquad\qquad E^\circ_{\text{cathode}} = 1.229$ V

$\quad\;\; -\;\; 2H_2O(l) + 2e^- \rightleftharpoons H_2(g) + 2\,OH^- \qquad\qquad E^\circ_{\text{anode}} = -0.828$ V

$\overline{\quad\;\; \tfrac{1}{2}O_2(g) + H_2(g) + 2H^+ + 2\,OH^- \rightleftharpoons 2H_2O(l) \qquad E^\circ = 2.057\text{ V}}$

$E_{\text{cathode}} = \left(1.229 - \dfrac{0.05916}{2}\log \dfrac{1}{P_{O_2}^{1/2}[H^+]^2}\right)$

$\qquad\quad = \left(1.229 - \dfrac{0.05916}{2}\log \dfrac{1}{1^{1/2}[1]^2}\right) = 1.229$ V

$E_{\text{anode}} = \left(-0.828 - \dfrac{0.05916}{2}\log P_{H_2}[OH^-]^2\right)$

$\qquad\quad = \left(-0.828 - \dfrac{0.05916}{2}\log\log 1\,[1]^2\right) = -0.828$ V

$E_{\text{cell}} = E_{\text{cathode}} - E_{\text{anode}} = 1.229 - (-0.828) = 2.057$ V

(b) $K = \dfrac{1}{P_{O_2}^{1/2}\,P_{H_2}[H^+]^2\,[OH^-]^2}$

$\qquad = 10^{2E^\circ/0.05916} = 10^{2(2.057)/0.05916} = 3.5 \times 10^{69}$

(c) If the current is 10.0 A, the electron flow is $(10.0\text{ C/s})/(9.649 \times 10^4\text{ C/mol})$ $= 1.036 \times 10^{-4}$ mol e^-/s. Each mole of H_2 gives 2 moles of electrons, so the rate of H_2 reaction is $\tfrac{1}{2}(1.036 \times 10^{-4}) = 5.18 \times 10^{-5}$ mol H_2/s = 1.044×10^{-4} g H_2/s = 1.044×10^{-7} kg H_2/s. Time required to consume 1.00 kg $H_2 = 1/(1.044 \times 10^{-7}\text{ kg }H_2/s) = 9.58 \times 10^6$ s.

Days $= (9.58 \times 10^6\text{ s})\left(\dfrac{1\text{ h}}{3\,600\text{ s}}\right)\left(\dfrac{1\text{ day}}{24\text{ h}}\right) = 111$ days

Electrode Potentials

For each mol H_2, ½ mole O_2 is consumed.

1.00 kg H_2 = 496 mol $\Rightarrow$ 248 mol O_2 = 7.94 kg O_2

14-16.
$$Mg(OH)_2(s) + 2e^- \rightleftharpoons Mg(s) + 2\,OH^- \qquad E_+^\circ = -2.690\text{ V}$$
$$-\ Mg^{2+} + 2e^- \rightleftharpoons Mg(s) \qquad E_-^\circ = -2.360\text{ V}$$
$$\overline{Mg(OH)_2(s) \rightleftharpoons Mg^{2+} + 2\,OH^-} \qquad E^\circ = -0.330\text{ V}$$

$$K_{sp} = 10^{2(-0.330)/0.05916} = 7 \times 10^{-12}$$

14-17.
$$Ca^{2+} + 2e^- \rightleftharpoons Ca(s) \qquad E_+^\circ = -2.868\text{ V}$$
$$-\ Ca(\text{acetate})^+ + 2e^- \rightleftharpoons Ca(s) + \text{acetate}^- \qquad E_-^\circ = -2.891\text{ V}$$
$$\overline{Ca^{2+} + \text{acetate}^- \rightleftharpoons Ca(\text{acetate})^+} \qquad E^\circ = 0.023\text{ V}$$

$$K_f = 10^{2(0.023)/0.05916} = 6.0$$

14-18. The activities of the solid reagents do not change until they are used up. The only aqueous species, OH^-, is created at the cathode and consumed in equal amounts at the anode, so its concentration remains constant in the cell. Therefore, none of the activities in the Nernst equation change during the life cycle of the cell (until something is used up).

14-19. Cell: $Au(s)\,|\,Fe(CN)_6^{4-}(aq),\,Fe(CN)_6^{3-}(aq)\,||\,Ag(S_2O_3)_2^{3-}(aq),\,S_2O_3^{2-}(aq)\,|\,Ag(s)$

(a) $Ag(S_2O_3)_2^{3-} + e^- \rightleftharpoons Ag(s) + 2\,S_2O_3^{2-}$ $\qquad E_+^\circ = 0.017\text{ V}$

$\quad-\ Fe(CN)_6^{3-} + e^- \rightleftharpoons Fe(CN)_6^{4-}$ $\qquad E_-^\circ = 0.356\text{ V}$

$\overline{Ag(S_2O_3)_2^{3-} + Fe(CN)_6^{4-} \rightleftharpoons Ag(s) + Fe(CN)_6^{3-} + 2\,S_2O_3^{2-}}$

$E^\circ = 0.017 - 0.356 = -0.339\text{ V} \Rightarrow K = 10^{1(-0.339)/0.05916} = 1.9 \times 10^{-6}$

(b) right half-cell: $E_+ = \left(0.017 - 0.05916\,\log\dfrac{[S_2O_3^{2-}]^2}{[Ag(S_2O_3)_2^{3-}]}\right)$

$\qquad\qquad\qquad = \left(0.017 - 0.05916\,\log\dfrac{[0.055]^2}{[1.8\times 10^{-3}]}\right) = 0.004\text{ V}$

left half-cell: $E_- = \left(0.356 - 0.05916\,\log\dfrac{[Fe(CN)_6^{4-}]}{[Fe(CN)_6^{3-}]}\right)$

$\qquad\qquad\qquad = \left(0.356 - 0.05916\,\log\dfrac{[1.3\times 10^{-3}]}{[4.9\times 10^{-3}]}\right) = 0.390\text{ V}$

$E = E_+ - E_- = 0.004 - 0.390 = -0.386\text{ V}$

(c) The potential of the left half-cell (0.390 V) is more positive than the potential of the right half-cell (0.004 V). Electrons travel through the circuit from Ag to Au. Ag(s) is oxidized to $Ag(S_2O_3)_2^{3-}$ in the spontaneous net cell reaction.

14-20. $Br_2(l) + 2e^- \rightleftharpoons 2Br^-$ $E_+^\circ = 1.078$ V
$-\underline{Br_2(aq) + 2e^- \rightleftharpoons 2Br^-}$ $E_-^\circ = 1.098$ V
$Br_2(l) \overset{K}{\rightleftharpoons} Br_2(aq)$ $E^\circ = -0.020$ V

$K = 10^{2(-0.020)/0.05916} = 0.21 = [Br_2(aq)] \Rightarrow [Br_2(aq)] = 0.21$ M

$(0.21 \text{ mol } Br_2/L)(159.81 \text{ g } Br_2/\text{mol } Br_2) = 34$ g Br/L

14-21. (a) $2IO_3^- + I^- + 12H^+ + 10e^- \rightleftharpoons I_3^- + 6H_2O$ $E_+^\circ = 1.210$ V
$-\underline{5I_3^- + 10e^- \rightleftharpoons 15I^-}$ $E_-^\circ = 0.535$ V
$2IO_3^- + 16I^- + 12H^+ \rightleftharpoons 6I_3^- + 6H_2O$ $E^\circ = 0.675$ V

(b) $E^\circ = 0.675$ V $\Rightarrow K = 10^{10E^\circ/0.05916} = 10^{114}$

(c) $E = \left(1.210 - \dfrac{0.05916}{10} \log \dfrac{[I_3^-]}{[IO_3^-]^2[I^-][H^+]^{12}}\right)$

$- \left(0.535 - \dfrac{0.05916}{10} \log \dfrac{[I^-]^{15}}{[I_3^-]^5}\right)$

$= \left(1.210 - \dfrac{0.05916}{10} \log \dfrac{[1.00 \times 10^{-4}]}{[0.0100]^2[0.0100][1.00 \times 10^{-6}]^{12}}\right)$

$- \left(0.535 - \dfrac{0.05916}{10} \log \dfrac{[0.0100]^{15}}{[1.00 \times 10^{-4}]^5}\right) = 0.772 - 0.594 = 0.178$ V

The potential for the first half-reaction is more positive (0.772 V) than that for the second half-reaction (0.594 V), so the reaction goes in the forward direction as written above.

(d) $\dfrac{[I_3^-]^6}{[IO_3^-]^2 [I^-]^{16} [H^+]^{12}} = 10^{114}$

Putting in the concentrations $[IO_3^-] = 0.0100$ M, $[I_3^-] = 1.00 \times 10^{-4}$ M, and $[I^-] = 0.0100$ M allows us to solve for $[H^+] = 3 \times 10^{-9}$ M $\Rightarrow$ pH = 8.5.

14-22. (a) $AgCl(s) + e^- \rightleftharpoons Ag(s) + Cl^-$
$Hg_2Cl_2(s) + 2e^- \rightleftharpoons 2Hg(l) + 2Cl^-$

(b) $E = E_+ - E_- = 0.241 - 0.197 = 0.044$ V

14-23. (a) $VO_2^+ + 2H^+ + e^- \rightleftharpoons VO^{2+} + H_2O \qquad E° = 1.001$ V

$$E = E° - 0.059\,16 \log\left(\frac{[VO^{2+}]}{[VO_2^+][H^+]^2}\right)$$

$$= 1.001 - 0.059\,16 \log\left(\frac{[0.050]}{[0.025][10^{-2.00}]^2}\right) = 0.747 \text{ V}$$

(b) $E = 0.747 - 0.241 = 0.506$ V

14-24. (a) In Figure 14-13, we see Ag | AgCl at +0.197 V from S.H.E. A potential of –0.111 V from Ag | AgCl will be 0.197 – 0.111 = 0.086 V from S.H.E.

(b) We see Ag | AgCl at –0.044 V from S.C.E. A potential of 0.023 V from Ag | AgCl will be –0.044 + 0.023 = –0.021 V from S.C.E.

(c) We see S.C.E. at +0.044 V from Ag | AgCl. A potential of –0.023 V from S.C.E will be +0.044 – 0.023 = +0.021 V from Ag | AgCl.

14-25. $E = E_+ - E_-$

$$E = \left(0.771 - 0.059\,16 \log\frac{[Fe^{2+}]}{[Fe^{3+}]}\right) - (0.241)$$

$$E = \left(0.771 - 0.059\,16 \log(2.5 \times 10^{-3})\right) - (0.241) = 0.684 \text{ V}$$

14-26. The reduction potentials tell us that it is more difficult to reduce LFe(III) to LFe(II) than it is to reduce free Fe^{3+} to Fe^{2+}. This implies that Fe(III) is stabilized more by desferrioxamine than is Fe(II). That is, the formation constant for the reaction L + Fe(III) $\rightleftharpoons$ LFe(III) is greater than the formation constant for the reaction L + Fe(II) $\rightleftharpoons$ LFe(II).

14-27. A few half-reactions we might think of are

$Cu^{2+}(aq) + 2e^- \rightleftharpoons Cu(s) \qquad E° = 0.339$ V
$Zn^{2+}(aq) + 2e^- \rightleftharpoons Zn(s) \qquad E° = -0.762$ V
$Cl_2(g) + 2e^- \rightleftharpoons 2Cl^-(aq) \qquad E° = 1.361$ V
$H_2O(l) + e^- \rightleftharpoons \frac{1}{2}H_2(g) + OH^-(aq) \qquad E° = -0.828$ V
$Na^+(aq) + e^- \rightleftharpoons Na(s) \qquad E° = -2.714$ V

Reactants in the voltaic pile are $Zn(s)$, $Cu(s)$, $H_2O(l)$, $Na^+(aq)$, and $Cl^-(aq)$. Zn, Cu, and Cl^- could be oxidized. H_2O and Na^+ could be reduced. I don't find a spontaneous reaction among these reactants under standard conditions. However, the least negative standard potential comes from oxidizing Zn and reducing H_2O: $E° = -0.828 - (-0.762) = -0.064$ V. Even though this reaction is unfavorable under standard conditions, low concentrations of the products Zn^{2+}, H_2 (g), and OH^- make the reaction spontaneous until enough products accumulate to make the cell voltage negative. For oxidation of Zn and reduction of H_2O in the brine layer, electrons flow out of the bottom metallic layer of the drawing and into the top layer. The outermost layers of Cu at the bottom of the pile and Zn at the top of the pile do not participate in the reaction except as metallic conductors. At the beginning of the reaction, O_2 from air dissolved in the brine could provides a very favorable reaction ($\frac{1}{2}O_2 + 2H^+(aq) + 2e^- \rightleftharpoons H_2O$, $E° = 1.229$ V) until the little bit of O_2 is consumed.

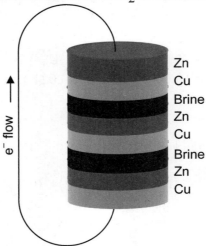

14-28. First we need to figure out how many moles of electrons are in 1 ampere·hour:

$$1 \text{ A·h} = \frac{(1 \text{ C/s})(3\,600 \text{ s/h})(1 \text{ h})}{9.649 \times 10^4 \text{ C/mol e}^-} = 0.037\,31 \text{ mol e}^-$$

In the lead-acid battery, each mole of Pb(s) transfers 2 mol e^- to each mole of PbO_2 to make $PbSO_4$. Therefore, 642.6 g of reactants transfers 2 mol e^-.

$$\frac{\text{ampere·hour}}{\text{kilogram of reactants}} = \frac{(2 \text{ mol e}^-)/(0.037\,31 \text{ mol e}^-/(\text{A·h}))}{0.642\,6 \text{ kg}} = 83.42 \frac{\text{A·h}}{\text{kg}}$$

In the fuel cell, each mole of O_2 receives 4 mol e^-.

$$\frac{(4 \text{ mol e}^-)/(0.037\,31 \text{ mol e}^-/(\text{A·h}))}{0.036\,031 \text{ kg}} = 2\,975 \frac{\text{A·h}}{\text{kg}}$$

CHAPTER 15
ELECTRODE MEASUREMENTS

15-1. (a) $Cu^{2+} + 2e^- \rightleftharpoons Cu(s)$ $\quad E_+^\circ = 0.339$ V

(b) $E_+ = 0.339 - \dfrac{0.059\,16}{2} \log \dfrac{1}{[Cu^{2+}]} = 0.309$ V

(c) $E = E_+ - E_- = 0.309 - 0.197 = 0.112$ V

15-2. (a) $Br_2(aq) + 2e^- \rightleftharpoons 2Br^-$ $\quad E_+^\circ = 1.098$ V

$E_+ = E_+^\circ - \dfrac{0.059\,16}{2} \log \dfrac{[Br^-]^2}{[Br_2(aq)]} = 1.098 - \dfrac{0.059\,16}{2} \log \dfrac{[0.234]^2}{[0.002\,17]} = 1.057$ V

(b) $E = E_+ - E_- = 1.057 - 0.241 = 0.816$ V

15-3. $SCN^- + Ag^+ \rightarrow AgSCN(s) \quad K = 1/K_{sp} = 1/1.1 \times 10^{-12}; \; V_e = 25.0$ mL

0.1 mL: Initial $SCN^- = (50.0\text{ mL})(0.100\text{ M}) = 5.00$ mmol

Added $Ag^+ = (0.1\text{ mL})(0.200\text{ M}) = 0.02$ mmol

$[SCN^-] = \dfrac{(5.00 - 0.02)\text{ mmol}}{50.1\text{ mL}} = 0.099\,4$ M

$[Ag^+] = K_{sp}/[SCN^-] = (1.1 \times 10^{-12})/0.099\,4 = 1.1 \times 10^{-11}$ M

Using Equation 15-1, $E = 0.558 + (0.059\,16)\log[Ag^+] = -0.090$ V

10.0 mL: In a similar manner, we find $[SCN^-] = 0.050\,0$ M

$[Ag^+] = 2.2 \times 10^{-11}$ M; $E = -0.073$ V

25.0 mL: At the equivalence point, $AgSCN(s) \rightleftharpoons Ag^+ + SCN^-$
$\qquad\qquad\qquad\qquad\qquad\qquad\qquad\qquad\quad x \quad\;\; x$

$[Ag^+][SCN^-] = x^2 = K_{sp}$

$\Rightarrow [Ag^+] = x = \sqrt{K_{sp}} = \sqrt{1.1 \times 10^{-12}} = 1.0_5 \times 10^{-6}$ M

$E = 0.558 + (0.059\,16)\log[Ag^+] = 0.204$ V

30.0 mL: This is 5.0 mL past $V_e \Rightarrow [Ag^+] = \dfrac{(5.0\text{ mL})(0.200\text{ M})}{80.0\text{ mL}} = 0.012\,5$ M

$E = 0.558 + (0.059\,16)\log[Ag^+] = 0.445$ V

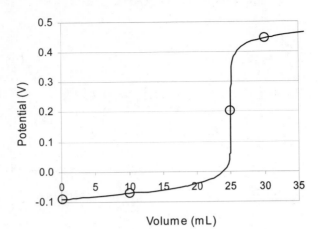

15-4. $Ag^+ + Br^- \rightarrow AgBr(s)$ $K = 1/K_{sp} = 1/5.0 \times 10^{-13}$; $V_e = 20.0$ mL

0.1 mL: Initial Ag^+ = (10.0 mL)(0.0500 M) = 0.500 mmol

Added Br^- = (0.1 mL)(0.0250 M) = 0.0025 mmol

$$[Ag^+] = \frac{(0.500 - 0.0025) \text{ mmol}}{10.1 \text{ mL}} = 0.0493 \text{ M}$$

Using Equation 15-1, $E = 0.558 + (0.05916)\log[Ag^+] = 0.481$ V

10.0 mL: In a similar manner, we find $[Ag^+] = 0.0125$ M

and $E = 0.445$ V

20.0 mL: $[Ag^+] = [Br^-] \Rightarrow [Ag^+] = \sqrt{K_{sp}} = \sqrt{5.0 \times 10^{-13}} = 7.0_7 \times 10^{-7}$ M

$E = 0.558 + (0.05916)\log[Ag^+] = 0.194$ V

30.0 mL: This volume is 10.0 mL past $V_e \Rightarrow [Br^-] = \dfrac{(10.0 \text{ mL})(0.0250 \text{ M})}{40.0 \text{ mL}}$

$= 0.00625$ M

$[Ag^+] = K_{sp}/[Br^-] = (5.0 \times 10^{-13})/0.00625 = 8.0 \times 10^{-11}$ M

$E = 0.558 + (0.05916)\log[Ag^+] = -0.039$ V

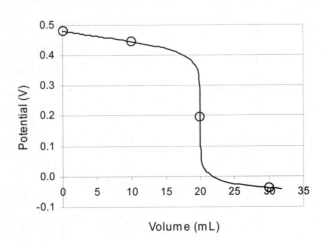

15-5. $Cl^- + Ag^+ \rightarrow AgCl(s)$ $K = 1/K_{sp} = 1/1.8 \times 10^{-10}$; $V_e = 50.0$ mL

1.0 mL: Initial Cl^- = (25.0 mL)(0.050 0 M) = 1.25 mmol
Added Ag^+ = (1.0 mL)(0.025 0 M) = 0.025 0 mmol

$$[Cl^-] = \frac{(1.25 - 0.025) \text{ mmol}}{26.0 \text{ mL}} = 0.047\ 1 \text{ M}$$

$[Ag^+] = K_{sp}/[Cl^-] = (1.8 \times 10^{-10})/0.047\ 1 = 3.8 \times 10^{-9}$ M

Using Equation 15-1, $E = 0.558 + (0.059\ 16) \log[Ag^+] = 0.060$ V

10.0 mL: In a similar manner, we find $[Cl^-] = 0.028\ 6$ M

$[Ag^+] = 6.3 \times 10^{-9}$ M; $E = 0.073$ V

50.0 mL: At the equivalence point, $AgCl(s) \rightleftharpoons Ag^+ + Cl^-$
$\ x x$

$[Ag^+][Cl^-] = x^2 = K_{sp}$

$\Rightarrow [Ag^+] = x = \sqrt{K_{sp}} = \sqrt{1.8 \times 10^{-10}} = 1.3_4 \times 10^{-5}$ M

$E = 0.558 + (0.059\ 16) \log[Ag^+] = 0.270$ V

60.0 mL: This volume is 10.0 mL past V_e

$\Rightarrow [Ag^+] = \dfrac{(10.0 \text{ mL})(0.025\ 0 \text{ M})}{85.0 \text{ mL}} = 0.002\ 9$ M

$E = 0.558 + (0.059\ 16) \log[Ag^+] = 0.408$ V

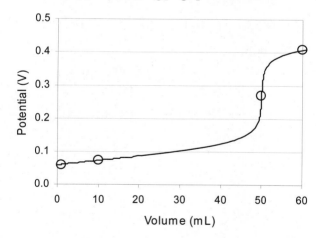

15-6. (a) $2Cl^- + Hg_2^{2+} \rightarrow Hg_2Cl_2(s)$ $K = 1/K_{sp} = 1/1.2 \times 10^{-18} = 8.3 \times 10^{17}$

$\underbrace{(2V_e \text{ (mL)})(0.100 \text{ M})}_{2 \times \text{mmol Hg}_2^{2+}} = \underbrace{(50.0 \text{ mL})(0.100 \text{ M})}_{\text{mmol Cl}^-} \Rightarrow V_e = 25.0$ mL

(b) Indicator electrode reaction: $Hg_2^{2+} + 2e^- \rightleftharpoons 2Hg(l)$ $E_+^\circ = 0.796$ V

Reference electrode: saturated calomel electrode $E_- = 0.241$ V

$$E = E_+ - E_- = \left(0.796 - \frac{0.05916}{2}\log\frac{1}{[Hg_2^{2+}]}\right) - (0.241)$$

$\underbrace{}_{\text{Potential of Hg}|Hg_2^{2+} \text{ indicator electrode}}$ $\quad\uparrow$ Constant potential of S.C.E. reference electrode

$$E = 0.555 + \frac{0.05916}{2}\log[Hg_2^{2+}] \qquad\qquad (A)$$

(c) 0.1 mL: Initial Cl$^-$ = (50.0 mL)(0.100 M) = 5.00 mmol

Added Hg_2^{2+} = (0.1 mL)(0.100 M) = 0.01 mmol

$$[Cl^-] = \frac{(5.00 - 2\times 0.01)\text{ mmol}}{50.1\text{ mL}} = 0.0994\text{ M}$$

$[Hg_2^{2+}] = K_{sp}/[Cl^-]^2 = (1.2\times 10^{-18})/(0.0994)^2 = 1.2_1\times 10^{-16}$ M

From Equation A, $E = 0.555 + (0.02958)\log[Hg_2^{2+}] = 0.084$ V

10.0 mL: In a similar manner, we find [Cl$^-$] = 0.0500 M

$[Hg_2^{2+}] = 4.8\times 10^{-16}$ M; $E = 0.102$ V

25.0 mL: At the equivalence point, $Hg_2Cl_2(s) \rightleftharpoons \underset{x}{Hg} + \underset{2x}{2Cl^-}$

$$[Hg_2^{2+}][Cl^-]^2 = x(2x)^2 = K_{sp}$$

$$\Rightarrow [Hg_2^{2+}] = x = \sqrt[3]{K_{sp}/4} = \sqrt[3]{1.2\times 10^{-18}/4} = 6.7\times 10^{-7}\text{ M}$$

$E = 0.555 + (0.02958)\log[6.7\times 10^{-7}] = 0.372$ V

30.0 mL: This volume is 5.0 mL past V_e.

$$\Rightarrow [Hg_2^{2+}] = \frac{(5.0\text{ mL})(0.100\text{ M})}{80.0\text{ mL}} = 0.00625\text{ M}$$

$E = 0.555 + (0.02958)\log[0.00625] = 0.490$ V

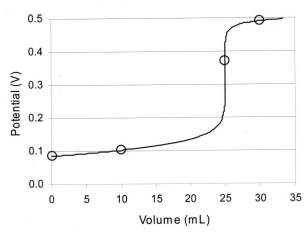

15-7. Relative mobilities:

$K^+ \rightarrow 7.62$ $\quad\quad\quad NO_3^- \rightarrow 7.40$

$5.19 \leftarrow Na^+$ $\quad\quad\quad 7.91 \leftarrow Cl^-$

K^+ diffuses faster than Na^+, thereby leading to excess positive charge on the right. Cl^- diffuses faster than NO_3^-, giving excess negative charge on the left.

Both the cation and anion diffusion cause negative charge to build up on the <u>left</u>.

15-8. Relative mobilities:

$H^+ \rightarrow 36.30$ $\quad\quad\quad Cl^- \rightarrow 7.91$

$7.62 \leftarrow K^+$ $\quad\quad\quad 7.91 \leftarrow Cl^-$

For 0.1 M HCl | 0.1 M KCl, the concentration of Cl^- is the same on both sides, so Cl^- contributes nothing to the junction potential. The mobility of H^+ is much greater than that of K^+, so H^+ diffuses to the right faster than K^+ diffuses to the left. The right side of the junction becomes positive.

For 0.1 M HCl | 3.5 M KCl, the positive sign of the junction potential tells us that the right side of the junction becomes positive. Although the mobility of H^+ is 5 times as great as the mobility of K^+, the concentration of K^+ is 35 times greater than that of H^+ on the left. The higher concentration of K^+ outweighs its lower mobility. If all else were equal, there would be a net diffusion of positive charge to the left. However, the concentration of Cl^- on the right side is 35 times greater than that on the left. Therefore, there will be a net diffusion of Cl^- from right to left, leaving excess positive charge on the right. The net result of the two competing effects is a small positive charge on the right.

15-9. The error shown in the graph is –0.33 pH units. The electrode will indicate $11.00 - 0.33 = 10.67$.

15-10. (a) The junction potential changes from –6.4 mV to –0.2 mV. A change of $6.4 - 0.2 = 6.2$ mV will appear to be a pH change of (6.2 mV)/(59.16 mV/pH unit) = +0.10 pH units.

(b) At pH 6.54, $[H^+] = 10^{-6.54} = 2.88 \times 10^{-7}$ M.
At pH 6.60, $[H^+] = 10^{-6.60} = 2.51 \times 10^{-7}$ M.

Percentage change in $[H^+]$ = $\dfrac{(2.88 - 2.51) \times 10^{-7}}{2.88 \times 10^{-7}} \times 100 = 13\%$

15-11. Measuring [H$^+$] with an ion-selective electrode is subject to the uncertain junction potential that plagues almost all direct potentiometric measurements. The unknown junction potential prevents us from measuring [H$^+$] exactly. On the other hand, the junction potential cannot change very much during small additions of titrant because the composition of the titration solution hardly changes. Therefore, changes in [H$^+$] can be measured with high precision even though the absolute value of [H$^+$] is not accurately known.

15-12. The liquid-based ion-selective electrode has a membrane containing a charged ion exchanger or a neutral ionophore that selectively binds analyte ion, which we suppose is the cation, C$^+$, for the sake of discussion. At the outer surface of the ion-selective membrane, there is an equilibrium between bound C$^+$ in the hydrophobic membrane and free C$^+$ in aqueous unknown solution. Free C$^+$ is hydrophilic, so it is not soluble in the hydrophobic membrane. A hydrophobic anion, R$^-$, is left behind in the membrane when the hydrophilic C$^+$ diffuses into the aqueous solution. On the inside surface of the membrane, there is a similar equilibrium between C$^+$ in the internal aqueous filling solution and bound C$^+$ in the membrane. If there is a difference between the concentration of C$^+$ in the external solution and the internal solution, there will be a net charge difference and therefore a potential difference across the membrane.

15-13. A compound electrode contains a second chemically active membrane outside the ion-selective membrane. The second membrane may be semipermeable and only allows the species of interest to pass through. Alternatively, the second membrane may contain a substance (such as an enzyme) that reacts with analyte to generate the species to which the ion-selective membrane responds.

15-14. The selectivity coefficient gives the relative response of an ion-selective electrode to the ion of interest and an interfering ion. The smaller the coefficient, the more selective is the electrode (smaller response to the interfering ion).

15-15. The electrode works the same way as described in Figure 15-6. The ion-selective, neutral ligand L is tridodecylamine (R$_3$N), which binds H$^+$ by acting as a base: R$_3$N + H$^+$ $\rightleftharpoons$ R$_3$NH$^+$. The hydrophobic anion is tetraphenylborate. pK_a for amines is normally so high (e.g., pK_a = 10.72 for (C$_2$H$_5$)$_3$N) that you would think that R$_3$N is fully protonated near pH 7.5 inside a living cell. However, amines dissolved in hydrophobic liquids are much harder to protonate than

Electrode Measurements

amines in aqueous solution, because the hydrophobic liquid does not stabilize ions as well as water does. Therefore, the amine remains mostly unprotonated until the aqueous phase in contact with the hydrophobic phase is very acidic (e.g., pH 3). Therefore, R_3N remains neutral in the ion-selective membrane and is available to serve as the H^+-binding ligand.

15-16. The conductive polymer translates an electric potential difference at the ion-exchange membrane into an electric potential at the metallic inner electrode. A liquid filling solution has the same function. An advantage of removing the filling solution is that there are no analyte ions in the conductive polymer to leak through the ion-selective membrane and thereby raise the detection limit. With no analyte ions leaking from the inside to the outside of the electrode, the detection limit of the electrode is lower.

15-17. $E_1 = \text{constant} + \dfrac{0.059\,16}{2} \log[1.00 \times 10^{-4}]$

$E_2 = \text{constant} + \dfrac{0.059\,16}{2} \log[1.00 \times 10^{-3}]$

$\Delta E = E_2 - E_1 = \dfrac{0.059\,16}{2} \log \dfrac{1.00 \times 10^{-3}}{1.00 \times 10^{-4}} = +0.029\,58 \text{ V}$

15-18. $[F^-]_{\text{Providence}} = 1.00 \text{ mg F}^-/\text{L} = 5.26 \times 10^{-5} \text{ M}$

$E_{\text{Providence}} = \text{constant} - 0.059\,16 \log[5.26 \times 10^{-5}]$

$E_{\text{Foxboro}} = \text{constant} - 0.059\,16 \log[F^-]_{\text{Foxboro}}$

$\Delta E = E_{\text{Foxboro}} - E_{\text{Providence}} = 0.0400 \text{ V}$

$= -0.059\,16 \log \dfrac{[F^-]_{\text{Foxboro}}}{5.26 \times 10^{-5}}$

$\Rightarrow [F^-]_{\text{Foxboro}} = 1.11 \times 10^{-5} \text{ M} = 0.211 \text{ mg/L}$

15-19. (a) $-0.230 = \text{constant} - (0.059\,16) \log(1.00 \times 10^{-3}) \Rightarrow \text{constant} = -0.407 \text{ V}$

(b) $-0.300 = -0.407 - (0.059\,16) \log x \Rightarrow x = 1.5_5 \times 10^{-2} \text{ M}$

15-20. Without sodium:

$-0.333 \text{ V} = \text{constant} + (0.059\,16) \log[3.44 \times 10^{-4}] \Rightarrow \text{constant} = -0.128 \text{ V}$

With sodium:

$E = -0.128 + (0.059\,16) \log[3.44 \times 10^{-4} + (5 \times 10^{-3})(0.100)] = -0.310 \text{ V}$

If we did not know that Na^+ was interfering, we would calculate the following concentration of Li^+ from the observed potential of -0.310 V:

$-0.310 \text{ V} = -0.128 + (0.059\,16) \log[Li^+] \Rightarrow [Li^+] = 8.4 \times 10^{-4} \text{ M}$

15-21. The potential without Mg^{2+} is

$$E = \text{constant} + \left(\frac{0.059\,16}{2}\right)\log[10^{-4}] = \text{constant} - 118.3 \text{ mV}$$

Addition of 1 mM Mg^{2+} gives an electrode potential of

$$E = \text{constant} + \left(\frac{0.059\,16}{2}\right)\log[0.000\,100 + (0.010)(0.001\,0)]$$

$$= \text{constant} - 117.1 \text{ mV}$$

The change is $-117.1 - (-118.3) = +1.2$ mV.

We could get the same +1.2 mV (= 0.001 2 V) increase by increasing $[Ca^{2+}]$:

$$E_{\text{initial}} = \text{constant} + \left(\frac{0.059\,16}{2}\right)\log[Ca^{2+}]_{\text{initial}}$$

$$E_{\text{final}} = \text{constant} + \left(\frac{0.059\,16}{2}\right)\log[Ca^{2+}]_{\text{final}}$$

$$E_{\text{final}} - E_{\text{initial}} = \left(\frac{0.059\,16}{2}\right)\log[Ca^{2+}]_{\text{final}} - \left(\frac{0.059\,16}{2}\right)\log[Ca^{2+}]_{\text{initial}}$$

$$0.001\,2 = \left(\frac{0.059\,16}{2}\right)\log\left(\frac{[Ca^{2+}]_{\text{final}}}{[Ca^{2+}]_{\text{initial}}}\right) \Rightarrow \frac{[Ca^{2+}]_{\text{final}}}{[Ca^{2+}]_{\text{initial}}} = 1.10$$

We could get the 1.2 mV increase in potential by increasing $[Ca^{2+}]$ by 10%.

15-22. (a) A graph of E (mV) versus $\log[NH_3 \text{ (M)}]$ gives a straight line whose equation is $E = 563.4 + 59.05 \log[NH_3]$. For $E = 339.3$ mV, $[NH_3] = 1.60 \times 10^{-4}$ M

(b) The sample analyzed contains $(100.0 \text{ mL})(1.60 \times 10^{-4} \text{ M}) = 0.0160$ mmol of nitrogen. But this sample represents just 2.00% (20.0 mL/1.00 L) of the food sample. Therefore, the food contains $0.016/0.0200 = 0.800$ mmol nitrogen = 11.2 mg N = 3.59 wt% nitrogen.

15-23. (a) For the nonactin-based electrode, K^+ has the largest selectivity coefficient, so it interferes the most. For the crown ether-based electrode, interference by Li^+ is greater than interference from K^+.

(b) From the diagram, I estimate $\log K^{\text{Pot}}_{NH_4^+, K^+}$ (nonactin) $= -1.0$ and $\log K^{\text{Pot}}_{NH_4^+, K^+}$ (crown ether) $= -1.5$.

$[K^{\text{Pot}}_{NH_4^+, K^+} \text{(crown ether)}]/[K^{\text{Pot}}_{NH_4^+, K^+} \text{(nonactin)}] = 10^{-1.5}/10^{-1.0} = 0.3$

15-24. (a) $E = \text{constant} + \beta \dfrac{0.05916}{3} \log[\text{La}^{3+}]_{\text{outside}}$

(b) If $[\text{La}^{3+}]$ increases by a factor of 10, the potential increases by $0.05916/3 = +19.7$ mV.

(c) $E_1 = \text{constant} + (1.00)\dfrac{0.05916}{3} \log(2.36 \times 10^{-4})$

$E_2 = \text{constant} + (1.00)\dfrac{0.05916}{3} \log(4.44 \times 10^{-3})$

$\Delta E = E_2 - E_1 = \dfrac{0.05916}{3} \log \dfrac{4.44 \times 10^{-3}}{2.36 \times 10^{-4}} = +25.1$ mV

(d) $E = \text{constant} + \dfrac{0.05916}{3} \log\left([\text{La}^{3+}] + \dfrac{1}{1\,200}[\text{Fe}^{3+}]\right)$

$0.100 = \text{constant} + \dfrac{0.05916}{3} \log(1.00 \times 10^{-4}) \Rightarrow \text{constant} = 0.178_9$ mV

$E = 0.178_9 + \dfrac{0.05916}{3} \log\left((1.00 \times 10^{-4}) + \dfrac{1}{1\,200}(0.010)\right) = +100.7$ mV

15-25.

	A	B	C	D
1	Least-Squares Spreadsheet			
2		x =		
3	Highlight cells B11:C13	log[Ca^{2+}]	y	
4	Type "= LINEST(C4:C8,	-4.471	-74.8	
5	B4:B8,TRUE,TRUE)	-3.471	-46.4	
6	For PC, press	-2.471	-18.7	
7	CTRL+SHIFT+ENTER	-1.471	10	
8	For Mac, press	-0.471	37.7	
9	COMMAND+RETURN			
10		LINEST output:		
11	m	28.1400	51.0939	b
12	s_m	0.0849	0.2416	s_b
13	R^2	1.0000	0.2683	s_y
14				
15	n =	5	B14 = COUNT(B4:B8)	
16	Mean y =	-18.44	B15 = AVERAGE(C4:C8)	
17	$\Sigma(x_i - \text{mean } x)^2 =$	10.00	B16 = DEVSQ(B4:B9)	
18				
19	Measured y =	-22.5	Input	
20	k = Number of replicate measurements of y =	1	Input	
21	Derived x =	-2.61528	B20 = (B19-C11)/B11	
22	s_x =	0.0105	B21 = (C13/B11)*SQRT((1/B20)+(1/B15)+((B19-B16)^2)/(B11^2*B17))	

Plot: $y = 28.14x + 51.094$, x-axis: $x = \log[\text{Ca}^{2+}]$

(a) The least-squares parameters are
$E = 51.09\ (\pm 0.24) + 28.14\ (\pm 0.08_5) \log[\text{Ca}^{2+}] \quad (s_y = 0.2_7)$

(b) Putting $E = -22.5$ into this equation gives
$\log[Ca^{2+}] = -2.615 \Rightarrow [Ca^{2+}] = 10^{-2.615} = 2.43 \times 10^{-3}$ M

(c) Cell B22 tells us that the uncertainty in $\log[Ca^{2+}]$ is $\pm 0.010\,5$, so the range for $[Ca^{2+}]$ is $10^{-2.604\,5}$ (= 2.49×10^{-3} M) to $10^{-2.625\,5}$ (= 2.37×10^{-3} M). A reasonable expression of uncertainty is $[Ca^{2+}] = 2.43\,(\pm 0.06) \times 10^{-3}$ M.

15-26. mean = 1.22_1; standard deviation = 0.05_0

$$\mu = \bar{x} \pm \frac{ts}{\sqrt{n}} = 1.22_1 \pm \frac{(2.17)(0.05_0)}{\sqrt{14}} = 1.22_1 \pm 0.02_9 = 1.19_2 \text{ to } 1.25_0$$

(Student's t was interpolated in Table 4-4 for 13 degrees of freedom.)

The value 1.19 is just barely outside the 95% confidence interval. It is so close to the 95% boundary that I would not say it is outside of "experimental error."

15-27. We will use the effective formation constant for PbY^{2-} (where Y = EDTA:

$Pb^{2+} + Y^{4-} \rightleftharpoons PbY^{2-}$ $K_f' = \alpha_{Y^{4-}} K_f = (1.5 \times 10^{-8})(10^{18.0}) = 1.5 \times 10^{10}$

$K_f' = \dfrac{[PbY^{2-}]}{[Pb^{2+}][EDTA]}$ (where EDTA is the total concentration of all species of EDTA not bound to Pb^{2+})

If we can find the concentrations of $[PbY^{2-}]$ and [EDTA], then we can compute the concentration of $[Pb^{2+}]$. There is a large excess of EDTA in the buffer. We expect essentially all lead to be in the form PbY^{2-}.

$[PbY^{2-}] = \dfrac{0.74 \text{ mL}}{100.74 \text{ mL}} (0.10 \text{ M}) = 7.3_5 \times 10^{-4}$ M

Total EDTA = $\dfrac{100.0 \text{ mL}}{100.74 \text{ mL}} (0.050 \text{ M}) = 0.049_6$ M

Free EDTA = $0.049_6 \text{ M} - \underbrace{7.3_5 \times 10^{-4} \text{ M}}_{\text{EDTA bound to } Pb^{2+}} = 0.048_9$ M

$K_f' = \dfrac{[PbY^{2-}]}{[Pb^{2+}][EDTA]}$

$\Rightarrow [Pb^{2+}] = \dfrac{[PbY^{2-}]}{K_f'[EDTA]} = \dfrac{7.3_5 \times 10^{-4}}{(1.5 \times 10^{10})(0.048_9)} = 1.0 \times 10^{-12}$ M

15-28. (a) We see from the flat portions of the curves that the higher the concentration of nitrite, the lower the electrode potential. pK_a for HNO_2 is 3.15. Below pH ≈ 4, a significant fraction of NO_2^- is converted to HNO_2, so the electrode senses less nitrite and the voltage rises.

(b) At high pH, there is a high concentration of OH⁻. The most likely explanation for the falling curves is that OH⁻ interferes with the electrode: The electrode responds to OH⁻ as if it were NO_2^-. The higher the pH, the greater the concentration of OH⁻, and the more the potential falls.

(c) The triangular region shows where the potential is independent of pH. The optimum pH appears to be near 4.5, where the widest range of nitrite can be measured without dependence on pH.

(d) Formula mass of nitrite (NO_2^-) is 46.006. 1 ppm ≈ 1 µg/mL = 21.74 µM

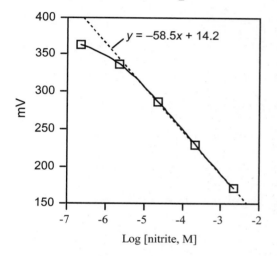

Here are the values I measured. Yours will be slightly different.

ppm	[NO_2^-] (µM)	mV
100	2 174	169.9
10	217.4	228.6
1	21.74	286.9
0.1	2.174	337.4
0.01	0.217 4	362.6

Data points deviate from linearity near log[NO_2^-] ≈ −5.5 or [NO_2^-] ≈ 3 µm.

CHAPTER 16
REDOX TITRATIONS

16-1.

$Ce^{4+} + e^- \rightleftharpoons Ce^{3+}$ $E° = 1.70$ V
$Fe^{3+} + e^- \rightleftharpoons Fe^{2+}$ $E° = 0.767$ V
──
$Ce^{4+} + Fe^{2+} \rightleftharpoons Ce^{3+} + Fe^{3+}$ $E° = 1.70 - 0.767 = 0.93_3$ V

$K = 10^{nE°/0.05916} = 10^{(1)(0.93_3)/0.05916} = 10^{15.77} = 6 \times 10^{15}$

Because there are only 2 significant digits in the exponent, $15._{77}$, we would be justified in writing the answer as 10^{16}.

16-2.
(a) $Ce^{4+} + Fe^{2+} \rightarrow Ce^{3+} + Fe^{3+}$

(b) $Fe^{3+} + e^- \rightleftharpoons Fe^{2+}$ $E° = 0.767$ V
    $Ce^{4+} + e^- \rightleftharpoons Ce^{3+}$ $E° = 1.70$ V

(c) $E = \left(0.767 - 0.05916 \log \frac{[Fe^{2+}]}{[Fe^{3+}]}\right) - (0.241)$ (A)

    $E = \left(1.70 - 0.05916 \log \frac{[Ce^{3+}]}{[Ce^{4+}]}\right) - (0.241)$ (B)

(d) 10.0 mL: Use Equation (A) with $[Fe^{2+}]/[Fe^{3+}] = 40.0/10.0$, because $V_e = 50.0$ mL $\Rightarrow E = 0.490$ V.

    25.0 mL: $[Fe^{2+}]/[Fe^{3+}] = 25.0/25.0 \Rightarrow E = 0.526$ V
    49.0 mL: $[Fe^{2+}]/[Fe^{3+}] = 1.0/49.0 \Rightarrow E = 0.626$ V
    50.0 mL: This is V_e, where $[Ce^{3+}] = [Fe^{3+}]$ and $[Ce^{4+}] = [Fe^{2+}]$. Equation 16-11 gives $E_+ = 1.23$ V and $E = 0.99$ V.
    51.0 mL: Use Equation B with $[Ce^{3+}]/[Ce^{4+}] = 50.0/1.0 \Rightarrow E = 1.36$ V.
    60.0 mL: $[Ce^{3+}]/[Ce^{4+}] = 50.0/10.0 \Rightarrow E = 1.42$ V
    100.0 mL: $[Ce^{3+}]/[Ce^{4+}] = 50.0/50.0 \Rightarrow E = 1.46$ V

16-3.
(a) $Ce^{4+} + Cu^+ \rightarrow Ce^{3+} + Cu^{2+}$

(b) $Ce^{4+} + e^- \rightleftharpoons Ce^{3+}$ $E° = 1.70$ V
    $Cu^{2+} + e^- \rightleftharpoons Cu^+$ $E° = 0.161$ V

(c) $E = \left(1.70 - 0.05916 \log \frac{[Ce^{3+}]}{[Ce^{4+}]}\right) - (0.197)$ (A)

    $E = \left(0.161 - 0.05916 \log \frac{[Cu^+]}{[Cu^{2+}]}\right) - (0.197)$ (B)

(d) 1.00 mL: Use Equation A with $[Ce^{3+}]/[Ce^{4+}] = 1.00/24.0$, because $V_e = 25.0$ mL $\Rightarrow E = 1.58$ V.

    12.5 mL: $[Ce^{3+}]/[Ce^{4+}] = 12.5/12.5 \Rightarrow E = 1.50$ V
    24.5 mL: $[Ce^{3+}]/[Ce^{4+}] = 24.5/0.5 \Rightarrow E = 1.40$ V

25.0 mL: For the equivalence point, add the two Nernst equations:

$$E_+ = 1.70 - 0.05916 \log \frac{[Ce^{3+}]}{[Ce^{4+}]}$$

$$E_+ = 0.161 - 0.05916 \log \frac{[Cu^+]}{[Cu^{2+}]}$$

$$2E_+ = 1.86_1 - 0.05916 \log \frac{[Ce^{3+}][Cu^+]}{[Ce^{4+}][Cu^{2+}]}$$

At the equivalence point, $[Ce^{3+}] = [Cu^{2+}]$ and $[Ce^{4+}] = [Cu^+]$. Therefore, the log term is zero, and $E_+ = 1.86_1/2 = 0.930$ V.

$$E = 0.930 - 0.197 = 0.733 \text{ V}$$

25.5 mL: We are 0.5 mL past the equivalence point, so the quotient $[Cu^+]/[Cu^{2+}]$ is 0.5/25.0. We put this value into Equation B to get $E = 0.065$ V.

30.0 mL: $[Cu^+]/[Cu^{2+}] = 5.0/25.0 \Rightarrow E = 0.005$ V

50.0 mL: $[Cu^+]/[Cu^{2+}] = 25.0/25.0 \Rightarrow E = -0.036$ V

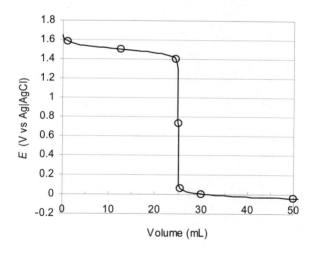

(e) The equivalence point potential vs S.H.E. is 0.930 V. Diphenylbenzidine sulfonic acid (violet → colorless) or diphenylamine sulfonic acid (red-violet → colorless) would be suitable. The change in potential near V_e is so great for this titration that other indicators such as diphenylamine (violet → colorless) or tris(2,2'-bipyridine)iron (pale blue → red) would also work.

16-4. (a) $Sn^{2+} + Tl^{3+} \rightarrow Sn^{4+} + Tl^+$

(b) $Sn^{4+} + 2e^- \rightleftharpoons Sn^{2+}$ $E° = 0.139$ V

$Tl^{3+} + 2e^- \rightleftharpoons Tl^+$ $E° = 0.77$ V

(c) $E = \left(0.139 - \dfrac{0.059\,16}{2} \log \dfrac{[Sn^{2+}]}{[Sn^{4+}]}\right) - (0.241)$ \hfill (A)

$E = \left(0.77 - \dfrac{0.059\,16}{2} \log \dfrac{[Tl^{+}]}{[Tl^{3+}]}\right) - (0.241)$ \hfill (B)

(d) 1.00 mL: Use Equation A with $[Sn^{2+}]/[Sn^{4+}] = 4.00/1.00$, because $V_e = 5.00$ mL $\Rightarrow E = -0.120$ V.

2.50 mL: $[Sn^{2+}]/[Sn^{4+}] = 2.50/2.50 \Rightarrow E = -0.102$ V

4.90 mL: $[Sn^{2+}]/[Sn^{4+}] = 0.10/4.90 \Rightarrow E = -0.052$ V

5.00 mL:
$$E_{+} = 0.139 - \dfrac{0.059\,16}{2} \log \dfrac{[Sn^{2+}]}{[Sn^{4+}]}$$

$$E_{+} = 0.77 - \dfrac{0.059\,16}{2} \log \dfrac{[Tl^{+}]}{[Tl^{3+}]}$$

$$2E_{+} = 0.90_9 - \dfrac{0.059\,16}{2} \log \dfrac{[Sn^{2+}][Tl^{+}]}{[Sn^{4+}][Tl^{3+}]}$$

At the equivalence point, $[Sn^{4+}] = [Tl^{+}]$ and $[Sn^{2+}] = [Tl^{3+}]$. Therefore, the log term is zero, and $E_{+} = 0.90_9/2 = 0.45_4$ V.

$E = 0.45_4 - 0.241 = 0.21$ V

5.10 mL: Use Equation B with $[Tl^{+}]/[Tl^{3+}] = 5.00/0.10 \Rightarrow E = 0.48$ V

10.0 mL: Use Equation B with $[Tl^{+}]/[Tl^{3+}] = 5.00/5.00 \Rightarrow E = 0.53$ V

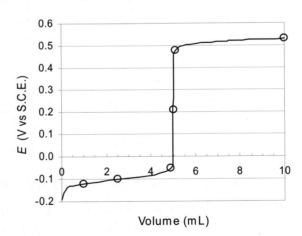

(e) The equivalence point potential vs S.H.E. is 0.45 V. Methylene blue (colorless → blue) would be suitable.

16-5. Titration reaction: $MnO_4^- + 5Fe^{2+} + 8H^+ \rightarrow Mn^{2+} + 5Fe^{3+} + 4H_2O$

$V_e = 15.0$ mL

$Fe^{3+} + e^- \rightleftharpoons Fe^{2+}$ $\qquad E° = 0.68$ V in 1 M H_2SO_4

$MnO_4^- + 8H^+ + 5e^- \rightleftharpoons Mn^{2+} + 4H_2O$ $\qquad E° = 1.507$ V

1.0 mL: $\quad E = E_+ - E_- = \left(0.68 - 0.05916 \log \frac{[Fe^{2+}]}{[Fe^{3+}]}\right) - 0.241$

$[Fe^{2+}]/[Fe^{3+}] = 14.0/1.0 \Rightarrow E = 0.371$ V

7.5 mL: $\quad [Fe^{2+}]/[Fe^{3+}] = 7.5/7.5 \Rightarrow E = 0.439$ V

14.0 mL: $\quad [Fe^{2+}]/[Fe^{3+}] = 1.0/14.0 \Rightarrow E = 0.507$ V

15.0 mL: At the equivalence point, use Equation E of Demonstration 16-1:

$6E_+ = 8.215 - 0.05916 \log \frac{1}{[H^+]^8}$. Using $[H^+] = 1$ gives $E_+ = 1.369$ V. $\quad E = E_+ - E_- = 1.369 - 0.241 = 1.128$ V

16.0 mL: $\quad E = E_+ - E_- = \left(1.507 - \frac{0.05916}{5} \log \frac{[Mn^{2+}]}{[MnO_4^-][H^+]^8}\right) - 0.241$

$[Mn^{2+}]/[MnO_4^-] = 15.0/1.0$ and $[H^+] = 1 \Rightarrow E = 1.252$ V

30.0 mL: $\quad [Mn^{2+}]/[MnO_4^-] = 15.0/15.0$ and $[H^+] = 1 \Rightarrow E = 1.266$ V

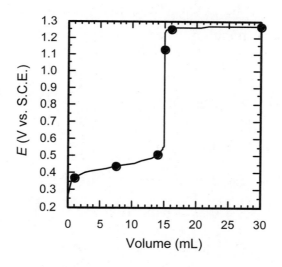

16-6. (a) Titration reaction: $Sn^{2+} + 2Fe^{3+} \rightarrow Sn^{4+} + 2Fe^{2+}$ $\qquad V_e = 25.0$ mL

(b) $Fe^{3+} + e^- \rightleftharpoons Fe^{2+}$ $\qquad E° = 0.732$ V

$Sn^{4+} + 2e^- \rightleftharpoons Sn^{2+}$ $\qquad E° = 0.139$ V

(c) $E = \left(0.732 - 0.05916 \log \frac{[Fe^{2+}]}{[Fe^{3+}]}\right) - (0.241)$ $\qquad$ (A)

$E = \left(0.139 - \frac{0.05916}{2} \log \frac{[Sn^{2+}]}{[Sn^{4+}]}\right) - (0.241)$ $\qquad$ (B)

(d) Representative calculations:

1.0 mL: $E_+ = 0.139 - \dfrac{0.05916}{2} \log \dfrac{[Sn^{2+}]}{[Sn^{4+}]}$

initial mol $Sn^{2+} = (25.0 \text{ mL})\left(0.0500 \dfrac{\text{mmol}}{\text{mL}}\right) = 1.25$ mmol

mol Fe^{3+} added $= (1.0 \text{ mL})\left(0.100 \dfrac{\text{mmol}}{\text{mL}}\right) = 0.10$ mmol

$[Sn^{4+}] = \dfrac{\frac{1}{2}(0.10 \text{ mmol})}{26.0 \text{ mL}} = 1.9_2 \times 10^{-3}$ M

$[Sn^{2+}] = \dfrac{1.25 - \frac{1}{2}(0.10) \text{ mmol}}{26.0 \text{ mL}} = 4.62 \times 10^{-2}$ M

$E_+ = 0.139 - \dfrac{0.05916}{2} \log \dfrac{[Sn^{2+}]}{[Sn^{4+}]}$

$E_+ = 0.139 - \dfrac{0.05916}{2} \log \dfrac{[4.62 \times 10^{-2}]}{[1.9_2 \times 10^{-3}]} = 0.098$ V

$E = E_+ - E_- = 0.098 - 0.241 = -0.143$ V

25.0 mL: At V_e, we add the two indicator electrode Nernst equations. To make the factor in front of the log term the same in both equations, we can multiply the $Sn^{4+} \mid Sn^{2+}$ equation by 2:

$E_+ = 0.139 - \dfrac{0.05916}{2} \log \dfrac{[Sn^{2+}]}{[Sn^{4+}]}$

$2E_+ = 0.278 - 0.05916 \log \dfrac{[Sn^{2+}]}{[Sn^{4+}]}$

$E_+ = 0.732 - 0.05916 \log \dfrac{[Fe^{2+}]}{[Fe^{3+}]}$

Now add the last two equations to get

$3E_+ = 1.010 - 0.05916 \log \left(\dfrac{[Sn^{2+}][Fe^{2+}]}{[Sn^{4+}][Fe^{3+}]}\right)$

But at the equivalence point, $2[Sn^{4+}] = [Fe^{2+}]$ and $2[Sn^{2+}] = [Fe^{3+}]$. Substituting these identities into the log term gives

$3E_+ = 1.010 - 0.05916 \log \left(\dfrac{[Sn^{2+}]2[Sn^{4+}]}{[Sn^{4+}]2[Sn^{2+}]}\right)$

So the log quotient in the log term is 1 and the $\log(1) = 0$. Therefore $E_+ = 1.010/3 = 0.337$ V and $E = E_+ - E_- = 0.337 - 0.241 = 0.096$ V.

26.0 mL: $E_+ = 0.732 - 0.05916 \log \dfrac{[Fe^{2+}]}{[Fe^{3+}]}$

Redox Titrations

There is 1.0 mL of Fe^{3+} beyond the equivalence point.

$$[Fe^{3+}] = \frac{(1.0 \text{ mL})(0.100 \text{ M})}{51.0 \text{ mL}} = 1.9_6 \times 10^{-3} \text{ M}$$

The first 25.0 mL of Fe^{3+} were converted to Fe^{2+}, so

$$[Fe^{2+}] = \frac{(25.0 \text{ mL})(0.100 \text{ M})}{51.0 \text{ mL}} = 4.90 \times 10^{-2} \text{ M}$$

$$E_+ = 0.732 - 0.059\,16 \log \frac{[Fe^{2+}]}{[Fe^{3+}]}$$

$$E_+ = 0.732 - 0.059\,16 \log \frac{[4.90 \times 10^{-2}]}{[1.9_6 \times 10^{-3}]} = 0.649 \text{ V}$$

$$E = E_+ - E_- = 0.649 - 0.241 = 0.408 \text{ V}$$

mL	E (V)	mL	E (V)	mL	E (V)
1.0	−0.143	24.0	−0.061	26.0	0.408
12.5	−0.102	25.0	0.096	30.0	0.450

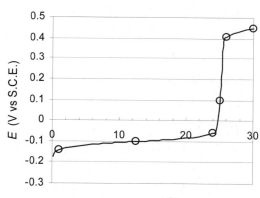

16-7. (a) $2Fe^{3+}$ + ascorbic acid + $H_2O \rightarrow 2Fe^{2+}$ + dehydroascorbic acid + $2H^+$

(b) The equivalence volume is 10.0 mL.

5.0 mL: Half of the Fe^{3+} is titrated, and the ratio $[Fe^{2+}]/[Fe^{3+}]$ is 5.0/5.0:

$$Fe^{3+} + e^- \rightleftharpoons Fe^{2+} \quad E° = 0.767 \text{ V}$$

$$E = E_+ - E_- = \left(0.767 - 0.059\,16 \log \frac{[Fe^{2+}]}{[Fe^{3+}]}\right) - (0.197)$$

$$= \left(0.767 - 0.059\,16 \log \frac{5.0}{5.0}\right) - (0.197) = 0.570$$

10.0 mL: This is the equivalence point. We multiply the ascorbic acid Nernst equation by 2 and add it to the iron Nernst equation. The reason for multiplying by 2 is to cancel the 2 in the denominator of

(0.059 16)/2 in front of the log term. When the factors in front of the log terms are equal to each other, we can add the log terms.

$$E_+ = 0.767 - 0.059\,16 \log \frac{[Fe^{2+}]}{[Fe^{3+}]}$$

$$2E_+ = 2\left(0.390 - \frac{0.059\,16}{2} \log \frac{[\text{ascorbic acid}]}{[\text{dehydro}][H^+]^2}\right)$$

$$3E_+ = 1.547 - 0.059\,16 \log \frac{[Fe^{2+}][\text{ascorbic acid}]}{[Fe^{3+}][\text{dehydro}][H^+]^2}$$

The stoichiometry of the titration reaction tells us that $[Fe^{2+}] = 2[\text{dehydroascorbic acid}]$ and $[Fe^{3+}] = 2[\text{ascorbic acid}]$. Inserting these equalities into the log term above gives

$$3E_+ = 1.547 - 0.059\,16 \log \frac{2[\text{dehydro}]\;[\text{ascorbic acid}]}{2[\text{ascorbic acid}]\,[\text{dehydro}][H^+]^2}$$

$$3E_+ = 1.547 - 0.059\,16 \log \frac{1}{[H^+]^2}$$

Using $[H^+] = 10^{-0.30}$ gives $E_+ = 0.504$ V and

$$E = 0.504 - 0.197 = 0.307 \text{ V}$$

15.0 mL: The ratio [dehydro]/[ascorbic acid] is 10.0/5.0.

dehydroascorbic acid + $2H^+$ + $2e^-$ ⇌ ascorbic acid + H_2O

$E° = 0.390$ V

$$E = E_+ - E_- = \left(0.390 - \frac{0.059\,16}{2} \log \frac{[\text{ascorbic acid}]}{[\text{dehydro}][H^+]^2}\right) - (0.197)$$

$$= \left(0.390 - \frac{0.059\,16}{2} \log \frac{[5.0]}{[10.0][10^{-0.30}]^2}\right) - (0.197) = 0.184 \text{ V}$$

16-8. For the solid titration curve, tris(2,2'-bipyridine)iron (red → pale blue) or tris(5-nitro-1,10-phenanthroline)iron (red-violet → pale blue) or tris(2,2'-bipyridine)ruthenium (yellow → pale blue) would be suitable. For the dashed curve, diphenylamine sulfonic acid (colorless → red violet) or diphenylbenzidine sulfonic acid (colorless → violet) would be suitable. The first hint of color would be taken as the end point.

16-9. The reduction potentials are

$Sn^{4+} + 2e^- \rightleftharpoons Sn^{2+}$ $\qquad E° = 0.139$ V in 1 M HCl

$Mn(EDTA)^- + e^- \rightleftharpoons Mn(EDTA)^{2-}$ $\qquad E° = 0.825$ V

The end point will be between 0.139 and 0.825 V. Tris(2,2'-bipyridine) iron has too high a reduction potential (1.120 V) to be useful for this titration.

16-10. I^- reacts with I_2 to give I_3^-, thereby increasing the solubility of I_2 and decreasing its volatility.

16-11. In disproportionation, an element in one oxidation state is converted into the same element in two different oxidation states. The oxidation state of iodine in each compound is: I_2 (0), HOI(+1), IO_3^-(+5), I^-(–1). Balanced half reactions are

$$I_2 + 2H_2O \rightarrow 2HOI + 2e^- + 2H^+ \quad (A)$$

$$I_2 + 6H_2O \rightarrow 2IO_3^- + 10e^- + 12H^+ \quad (B)$$

$$I_2 + 2e^- \rightarrow 2I^- \quad (C)$$

Reactions A and C can be combined to give

$$2I_2 + 2H_2O \rightarrow 2HOI + 2I^- + 2H^+ \quad (D)$$

Reactions B and C can be combined to give

$$6I_2 + 6H_2O \rightarrow 2IO_3^- + 10I^- + 12H^+ \quad (E)$$

Reactions D and E are each balanced. They can occur in any combination, ranging from all of one to all of the other.

16-12. (a) KI solution is colorless. After reaction with O_3 to create I_3^-, the solution would be pale red.

(b) 29.33 mL of 0.050 44 M $S_2O_3^{2-}$ = 1.479_4 mmol $S_2O_3^{2-}$. The reaction with iodine is $I_3^- + 2S_2O_3^{2-} \rightarrow 3I^- + S_4O_6^{2-}$, so there must have been $\frac{1}{2}(1.479_4$ mmol$) = 0.739_7_0$ mmol I_3^-. Because 1 mol O_3 generates 1 mol I_3^-, the bulb contained 0.739_7_0 mmol O_3. With a formula mass of 47.998 g O_3/mol, 0.739_7_0 mmol O_3 = 35.50 mg O_3.

(c) It does matter. I_3^- is present from the beginning of the thiosulfate titration, so starch should not be added until immediately before the end point.

16-13. Each mole of NH_3 liberated in the Kjeldahl digestion reacts with one mole of H^+ in the standard H_2SO_4 solution. Six moles of H^+ left (3 mol H_2SO_4) after reaction with NH_3 will react with 1 mol iodate by Reaction 16-20 to release 3 mol I_3^-. Two moles of thiosulfate react with one mole of I_3^- in Reaction 16-21. Therefore, each mole of thiosulfate corresponds to one-half mole of residual H_2SO_4.

$$\text{mol NH}_3 = 2(\text{initial mol H}_2\text{SO}_4 - \text{final mol H}_2\text{SO}_4)$$

$$\text{mol NH}_3 = 2(\text{initial mol H}_2\text{SO}_4 - \frac{1}{2} \times \text{mol thiosulfate})$$

16-14. (a) $IO_3^- + 8I^- + 6H^+ \rightleftharpoons 3I_3^- + 3H_2O$. The stock solution contained $\{0.804\,3$ g KIO_3 (FM 214.00) + 6.0 g KI (FM 166.00)$\}$/100 mL, which translates into 0.037 58 M KIO_3 plus 0.36 M KI. The mole ratio KI/KIO_3 = 9.6, which is excess of the 8:1 ratio required in the reaction. 5.00 mL of the stock solution contain 0.187 9_2 mmol KIO_3 plus 1.5 mmol KI. 1.0 mL of 6.0 M H_2SO_4 contains 6 mmol H_2SO_4, which is a large excess for the reaction. Neither KI nor H_2SO_4 needs to be measured accurately.

(b) $I_3^- + SO_3^{2-} + H_2O \rightarrow 3I^- + SO_4^{2-} + 2H^+$

(c) 0.187 9_2 mmol KIO_3 delivered to the wine generates $3 \times 0.187\,9_2 = 0.563\,7_6$ mmol I_3^-. The excess, unreacted I_3^- required 12.86 mL of 0.048 18 M $Na_2S_2O_3 = 0.619\,5_9$ mmol $Na_2S_2O_3$. Each mole of unreacted I_3^- requires two moles of $Na_2S_2O_3$, so there must have been $(0.619\,5_9)/2 = 0.309\,8$ mmol I_3^- left over from the reaction with sulfite. Therefore, the I_3^- that reacted with sulfite was $(0.563\,7_6 - 0.309\,8) = 0.254\,0$ mmol I_3^-. One mole of I_3^- reacts with one mole of sulfite, so there must have been 0.254 0 mmol SO_3^{2-} in 50.0 mL of wine. $[SO_3^{2-}] = 0.254\,0$ mmol/50.0 mL $= 5.079 \times 10^{-3}$ M. With a formula mass of 80.06 for sulfite, the sulfite content is 406.6 mg/L.

(d) $s_{pooled} = \sqrt{\dfrac{2.2^2(3-1) + 2.1^2(3-1)}{3+3-2}} = 2.15$

$t_{calculated} = \dfrac{|277.7 - 273.2|}{2.15}\sqrt{\dfrac{3 \cdot 3}{3+3}} = 2.56$

$t_{table} = 2.776$ for 95% confidence and $3 + 3 - 2 = 4$ degrees of freedom
$t_{calculated} < t_{table}$, so the difference is <u>not</u> significant at 95% confidence level.

16-15. (a) $I_2(aq) + 2e^- \rightleftharpoons 2I^-$ $E° = 0.620$ V

$\underline{I_3^- + 2e^- \rightleftharpoons 3I^-E° = 0.535\text{ V}}$

$I_2(aq) + I^- \rightleftharpoons I_3^-$ $E° = 0.620 - 0.535 = 0.085$ V

$K = 10^{E°/0.059\,16} = 10^{2(0.085)/0.059\,16} = 7 \times 10^2$

Redox Titrations

(b) $\quad I_2(s) + 2e^- \rightleftharpoons 2I^-$ $\qquad E° = 0.535$ V

$\underline{\quad I_3^- + 2e^- \rightleftharpoons 3I^- \qquad\qquad\qquad E° = 0.535 \text{ V} \quad}$

$\quad I_2(s) + I^- \rightleftharpoons I_3^-$ $\qquad E° = 0.535 - 0.535 = 0.000$ V

$\qquad K = 10^{2(-0.000)/0.059\,16} = 1.0$

(c) $\quad I_2(s) + 2e^- \rightleftharpoons 2I^-$ $\qquad E° = 0.535$ V

$\underline{\quad I_2(aq) + 2e^- \rightleftharpoons 2I^- \qquad\qquad E° = 0.620 \text{ V} \quad}$

$\quad I_2(s) \rightleftharpoons I_2(aq)$ $\qquad E° = 0.535 - 0.620 = -0.085$ V

$\qquad K = [I_2(aq)] = 10^{2(-0.085)/0.059\,16} = 1.3 \times 10^{-3}$ M $= 0.34$ g of I_2/L

16-16. (a) $n = \dfrac{PV}{RT} = \dfrac{200 \times 10^{-9} \text{ bar} \cdot 1 \text{ L}}{8.314 \times 10^{-2} \text{ L} \cdot \text{bar/(mol} \cdot \text{K)} \cdot 300 \text{ K}} \approx 8$ nmol O_3

(b) One mole of O_3 generates one mole of I_3^-, which requires two moles of thiosulfate for titration, and 8 nmol O_3 generates 8 nmol I_3^-, requiring 16 nmol $S_2O_3^{2-}$. If 8 nmol I_3^- were generated in, say, 10 mL of solution, the concentration would be 8×10^{-9} mol/(0.01 L) = 8×10^{-7} M. The text says that the color of $\sim 5 \times 10^{-7}$ M I_3^- is just barely discernible with a starch indicator. This amount would be too little to titrate. If the ozone were absorbed into 1 mL of solution instead of 10 mL, the concentration of I_3^- would be 8 μM, which ought to be enough to see and to allow a crude titration. If the thiosulfate titrant were delivered from a microliter pipet, perhaps a total volume of 100 μL (0.1 mL) might be used. The required concentration of titrant would be (16 nmol $S_2O_3^{2-}$)/(10^{-4} L) = 0.16 mM. This concentration is very low, but probably could be used. A low accuracy titration might just barely be feasible.

16-17. (a) $YBa_2Cu_3O_7$ contains 1 Cu^{3+} and 2 Cu^{2+}. $YBa_2Cu_3O_{6.5}$ contains no Cu^{3+} and 3 Cu^{2+}. The moles of Cu^{3+} in the formula $YBa_2Cu_3O_{7-z}$ are therefore $1 - 2z$. The moles of superconductor in 1 g of superconductor are $(1\text{g})/[(666.246 - 15.999\,4\,z)\text{g/mol}]$. The difference between experiments B and A is $5.68 - 4.55 = 1.13$ mmol $S_2O_3^{2-}$/g superconductor. Because 1 mol of thiosulfate is equivalent to 1 mol of Cu^{3+}, there are 1.13 mmol Cu^{3+}/g superconductor.

$$\frac{\text{mol Cu}^{3+}}{\text{mol superconductor}} = 1 - 2z = \frac{1.13 \times 10^{-3} \text{ mol Cu}^{3+}}{\left(\dfrac{1 \text{ g superconductor}}{(666.246 - 15.9994z) \text{ g/mol}}\right)}$$

Solving this equation gives $z = 0.125$. The formula is $YBa_2Cu_3O_{6.875}$.

(b) $$1 - 2z = \frac{[5.68\,(\pm 0.05) - 4.55\,(\pm 0.10)] \times 10^{-3}}{\left(\dfrac{1}{666.246 - 15.9994z}\right)}$$

$$1 - 2z = \frac{1.13\,(\pm 0.112) \times 10^{-3}}{\left(\dfrac{1}{666.246 - 15.9994z}\right)}$$

$1 - 2z = 0.752\,86\,(\pm 0.074\,49) - 0.018\,079\,(\pm 0.001\,789)z$

$0.247\,124\,(\pm 0.074\,488) = 1.981\,92\,(\pm 0.001\,79)z$

$z = 0.125 \pm 0.038$

The formula is $YBa_2Cu_3O_{6.875\,\pm 0.038}$.

CHAPTER 17
INSTRUMENTAL METHODS IN ELECTROCHEMISTRY

17-1. (a) In electrogravimetric analysis, analyte is quantitatively deposited on an electrode by an electrolysis reaction. The increase in mass of the electrode tells how much analyte was present.

(b) Three ways to tell that the electrodeposition is over: (1) Observe the disappearance of color in a solution from which a colored species is removed. (2) Expose most, but not all, of the surface of the working electrode to the solution during electrolysis. After a period of electrolysis, raise the beaker or add water so that fresh surface of the electrode is exposed to the solution. After an additional period of electrolysis, see if the newly exposed electrode surface has a deposit. If it does, repeat the procedure. If not, the electrolysis is done. (3) Remove a small sample of solution and perform a qualitative test for analyte.

17-2. Reaction 17-3a generates I_2 at an electrode. Two moles of electrons flow through the circuit for each mole of I_2 that is generated. If a constant current I (coulombs/second) flows for a time t (seconds), the moles of electrons passing through the circuit are It/nF, where n is 1 charge/electron and F is the Faraday constant (coulombs/mole). By knowing the time and current required to conduct the electrolysis Reaction 17-3a to reach the end point of Reaction 17-3b, we know how many moles of I_2 were necessary to react with the H_2S. Each mole of I_2 generated in Reaction 17-3a consumes one mole of H_2S in Reaction 17-3b.

17-3. Answer: V_2. V_2 is the potential difference between the working and reference electrodes, which is what we want to regulate in a controlled potential electrolysis. V_1 is the potential difference between the auxiliary and reference electrodes, which is uncontrolled and varies during the electrolysis.

17-4. The mercury drop is the working electrode at which the reaction of interest of the analyte occurs. Usually this reaction is reduction of analyte. The Pt auxiliary electrode is where the other redox reaction (usually an oxidation) occurs. Current flows between the working and auxiliary electrodes. The potential of the working electrode is controlled with respect to the potential of the calomel reference electrode, at which negligible current flows. The purpose of the reference electrode is to establish a stable potential against which the potential of the working electrode can be measured.

162   Chapter 17

17-5. Faradaic current arises from a redox reaction at the surface of an electrode. For example, analyte or water might be oxidized or reduced at the surface of an inert electrode or a Cu electrode might be oxidized to release Cu^{2+} into solution. Charging current is the migration of ions toward or away from an electrode as the potential changes. For example, as an electrode is driven more negative by the power supply, cations migrate toward the electrode and anions migrate away from the electrode. In the staircase voltage ramp, there is a large flow of ions toward or away from the working electrode each time a potential step is applied. Charging current rapidly subsides as the ionic atmosphere around the electrode equilibrates with the new potential. If we wait 1 s after applying the potential step, the charging current has died down to a small value. The faradaic current, which we want to measure, does not die off as rapidly as the charging current. After 1 s, the faradaic current can be measured with little interference from the charging current.

17-6. Mass of Cu(s) deposited = 16.414 − 15.327 = 1.087 g = $0.017\,10_6$ mol Cu
[Cu(II)] in unknown = $(0.017\,10_6$ mol)/(0.050 0 L) = 0.342 M

17-7. If the reagent contains only $CoCl_2$ and H_2O, we can write

mol Co in sample = $\left(\dfrac{\text{grams of Co deposited}}{\text{atomic mass of Co}}\right)$

mol $CoCl_2$ = mol Co

grams of $CoCl_2$ = $\left(\dfrac{\text{grams of Co deposited}}{\text{atomic mass of Co}}\right)$ (FM of $CoCl_2$) = 0.218 93 g

grams of $CoCl_2$ = $\left(\dfrac{0.099\,37\text{ g}}{58.933\,2\text{ g/mol}}\right)$(129.839 g/mol) = 0.218 93 g

grams of H_2O = 0.402 49 − 0.218 93 = 0.183 56 g

$\dfrac{\text{moles of }H_2O}{\text{moles of Co}} = \dfrac{0.183\,56/\text{FM of }H_2O}{0.099\,37/\text{atomic mass of Co}} = 6.043$

The reagent composition is close to $CoCl_2 \cdot 6H_2O$.

17-8. 75.00 mL of 0.023 80 M KSCN = 1.785 mmol SCN^-, which gives 1.785 mmol AgSCN, containing 0.103 7 g SCN^-.
Final mass = 12.463 8 + 0.103 7 = 12.567 5 g

17-9. $Pb(\text{lactate})_2 + 2H_2O \rightarrow PbO_2(s) + 2\text{ lactate}^- + 4H^+ + 2e^-$
    Pb^{2+} Pb^{4+}

Lead lactate is oxidized to PbO_2 at the anode. (The anode is where oxidation occurs.) Mass of lead lactate (FM 385.3) giving 0.111 1 g PbO_2 (FM = 239.2) is (385.3/239.2)(0.111 1 g) = 0.179 0 g.

wt% Pb = $\dfrac{0.179\ 0}{0.326\ 8} \times 100$ = 54.77 wt%

17-10. $2I^- \rightarrow I_2 + 2e^- \Rightarrow$ One mole of I_2 is created when two moles of electrons flow.

q = coulombs of e^- = $I\left(\dfrac{C}{s}\right) \times t\ (s)$ = (812 s)(52.6 × 10^{-3} C/s) = 42.7$_1$ C

N = mol e^- = $\dfrac{q}{nF}$ = $\dfrac{42.7_1\ C}{(1\ \text{charge/electron})(96\,485\ C/mol)}$ = 4.42$_7$ × 10^{-4} mol e^-

$(0.442\ 7\ \text{mmol}\ e^-)\left(\dfrac{1\ \text{mmol}\ I_2}{2\ \text{mmol}\ e^-}\right)$ = 0.221 3 mmol I_2

Therefore, there must have been 0.221 3 mmol H_2S (FM 34.08) = 7.542 mg H_2S/50.00 mL = 7.542 × 10^3 µg of H_2S/50.00 mL = 1.51 × 10^2 µg/mL.

17-11. (a) Cathode reaction at (–) electrode: $H_2O + e^- \rightarrow \frac{1}{2}H_2 + OH^-$

Anode reaction at (+) electrode: $H_2O \rightarrow \frac{1}{2}O_2 + 2H^+ + 2e^-$

(b) q = coulombs of e^- = $I\left(\dfrac{C}{s}\right) \times t\ (s)$ = (89.2 mA)(666 s)
= (89.2 × 10^{-3} C/s)(666 s) = 59.4$_1$ C

N = mol e^- = $\dfrac{q}{nF}$ = $\dfrac{59.4_1\ C}{(1\ \text{charge/electron})(96\,485\ C/mol)}$ = 0.615$_7$ mmol e^-

HA in unknown = (0.615$_7$ mmol)/(5.00 mL) = 0.123 M

17-12. (a) q = coulombs of e^- = It = 0.005 C/s × 0.1 s = 0.000 5 C

N = mol e^- = $\dfrac{q}{nF}$ = $\dfrac{0.000\ 5C}{(1\ \text{charge/electron})(96\,485\ C/mol)}$ = 5.$_2$ × 10^{-9} mol e^-

(b) A 0.01 M solution of a two-electron reductant delivers 0.02 moles of electrons/liter.

$\dfrac{5._2 \times 10^{-9}\ \text{moles}}{0.02\ \text{moles/liter}}$ = 2.$_6$ × 10^{-7} L = 0.000 2$_6$ mL

17-13. (a) $n = \dfrac{PV}{RT} = \dfrac{(0.996 \text{ bar})(0.049\,22 \text{ L})}{(0.083\,145 \text{ L bar K}^{-1} \text{ mol}^{-1})(303 \text{ K})} = 1.946$ mmol of H_2

(b) For every mole of H_2 produced, two moles of e^- are consumed and one mole of Cu is oxidized. Therefore, 1.946 mmol Cu^{2+} are produced and the concentration of EDTA is 1.946 mmol/47.36 mL = 0.041 09 M.

(c) 1.946 mmol of H_2 comes from 2(1.946) = 3.892 mmol of e^-
$= (3.892 \times 10^{-3}$ mol$)(96\,485$ C/mol$) = 3.755 \times 10^2$ C.
Time $= t = q/I = (3.755 \times 10^2$ C$)/(0.021\,96$ C/s$) = 1.710 \times 10^4$ s $= 4.750$ h

17-14. Trichloroacetate is reduced at −0.90 V and consumes $N = $ mol $e^- = \dfrac{q}{nF}$

$= \dfrac{224 \text{ C}}{(1 \text{ charge/electron})(96\,485 \text{ C/mol})} = 2.322$ mmol e^-.

Each Cl_3CCO_2H requires 2 electrons for reduction, so there must have been

$\left(\dfrac{1 \text{ mmol } Cl_3CCO_2H}{2 \text{ mmol } e^-}\right)(2.322 \text{ mmol } e^-) = 1.161$ mmol Cl_3CCO_2H (FM 163.39)

$= 0.189\,7$ g Cl_3CCO_2H.

For reduction of Cl_2HCCO_2H at −1.65 V, $N = $ mol $e^- = \dfrac{q}{nF}$

$= \dfrac{758 \text{ C}}{(1 \text{ charge/electron})(96\,485 \text{ C/mol})} = 7.856$ mmol e^-.

The total quantity of Cl_2HCCO_2H (FM 128.94) is

$\left(\dfrac{1 \text{ mmol } Cl_2HCCO_2H}{2 \text{ mmol } e^-}\right)(7.856 \text{ mmol } e^-) = 3.928$ mmol Cl_2HCCO_2H, of which 1.161 mmol came from reduction of Cl_3CCO_2H at −0.90 V.

Cl_2HCCO_2H in original sample = 3.928 mmol − 1.161 mmol
= 2.767 mmol = 0.356 8 g Cl_2HCCO_2H.

wt% trichloroacetic acid $= \dfrac{0.189\,7 \text{ g}}{0.721 \text{ g}} \times 100 = 26.3$ wt%

wt% dichloroacetic acid $= \dfrac{0.356\,8 \text{ g}}{0.721 \text{ g}} \times 100 = 49.5$ wt%

17-15. The corrected coulometric titration time is 387 − 6 = 381 s.
$q = It = (4.23$ mA$)(381$ s$) = 1.61_{16}$ C

$N = $ mol $e^- = \dfrac{q}{nF} = \dfrac{1.61_{16} \text{ C}}{(1 \text{ charge/electron})(96\,485 \text{ C/mol})} = 16.7$ μmol e^-

Instrumental Methods in Electrochemistry

Because one e^- is equivalent to one X^-, the concentration of organohalide is 16.7 µM. If all halogen is Cl, this quantity corresponds to 592 µg Cl/L.

17-16.
$$F = \frac{\text{coulombs}}{\text{mol}} = \frac{I \cdot t}{\text{mol}}$$

$$= \frac{[0.203\,639\,0\,(\pm 0.000\,000\,4)\text{ A}][18\,000.075\,(\pm 0.010)\text{ s}]}{[4.097\,900\,(\pm 0.000\,003)\text{ g}]/[107.868\,2\,(\pm 0.000\,2)\text{g/mol}]}$$

$$= \frac{[0.203\,639\,0\,(\pm 1.96 \times 10^{-4}\,\%)][18\,000.075\,(\pm 5.56 \times 10^{-5}\,\%)]}{[4.097\,900\,(\pm 7.32 \times 10^{-5}\,\%)]/[107.868\,2\,(\pm 1.85 \times 10^{-4}\,\%)]}$$

$$= 9.648\,667 \times 10^4\,(\pm 2.85 \times 10^{-4}\,\%) = 96\,486.6_7 \pm 0.2_8 \text{ C/mol}$$

17-17. (a) The Clark electrode has a Au-coated Pt cathode and a Ag|AgCl anode. Access to the electrode is controlled by a thin silicone rubber membrane through which O_2 from a test solution can diffuse. When the cathode is held at –0.75 V with respect to the anode, the following reactions occur:

Cathode: $O_2 + 4H^+ + 4e^- \rightarrow 2H_2O$

Anode: $4Ag + 4Cl^- \rightarrow 4AgCl + 4e^-$

The current is proportional to the concentration of dissolved O_2. The Ag guard electrode is kept at a negative potential to reduce any O_2 that diffuses in from the air through the top of the electrode. The electrode response is a current, not a voltage.

(b) A concentration of dissolved O_2 of "0.20 bar" means that dissolved O_2 is in equilibrium with 0.20 bar of O_2 in the atmosphere. From Henry's law,

$[O_2(aq)] = (0.001\,26 \text{ M/bar}) \times P_{O_2}$ (bar)

$= (0.001\,26 \text{ M/bar})(0.20 \text{ bar}) = 0.25 \text{ mM}$

17-18. (a) In the spreadsheet on the next page, columns B and C contain the standard addition volumes (V) and observed currents (I_{S+X}). Column D is the function $[S]_i(V_s/V_0)$ to plot on the x-axis, where $[S]_i$ is the concentration of standard prior to adding it to the sample, V_s is the volume of standard added, and V_0 is the initial volume of sample. Column E is the function $I_{S+X}(V/V_0)$ to plot on the y-axis. The negative x-intercept, 2.89 mM in cell B26, is the *original* concentration $[X]_i$ of ascorbic acid in the undiluted orange juice.

(b) Uncertainty in the intercept is estimated with the equation

$$\text{Standard deviation of } x\text{-intercept} = \frac{s_y}{|m|}\sqrt{\frac{1}{n} + \frac{\bar{y}^2}{m^2 \Sigma(x_i - \bar{x})^2}}$$

where s_y is the standard deviation of y, $|m|$ is the absolute value of the slope of the least-squares line, n is the number of data points ($n = 9$), $\bar{y}$ is the mean value of y for the 9 points, x_i are the individual values of x for the 9 points, and $\bar{x}$ is the mean value of x for the 9 points. Uncertainty computed in cell B33 is 0.098 mM. A reasonable expression of the concentration of ascorbic acid in the orange juice is 2.89 ± 0.10 mM.

	A	B	C	D	E
1	Vitamin C standard addition experiment				
2	Add 0.279 M ascorbic acid to 50.0 mL of orange juice				
3					
4		Vs =			
5	Vo (mL) =	mL ascorbic	I(s+x) =	x-axis function	y-axis function
6	50	acid added	signal (µA)	Si*Vs/Vo	I(s+x)*V/Vo
7	[S]i (mM) =	0.000	1.78	0.000	1.780
8	279	0.050	2.00	0.279	2.002
9		0.250	2.81	1.395	2.824
10		0.400	3.35	2.232	3.377
11		0.550	3.88	3.069	3.923
12		0.700	4.37	3.906	4.431
13		0.850	4.86	4.743	4.943
14		1.000	5.33	5.580	5.437
15		1.150	5.82	6.417	5.954
16				D7 = $A$8*B7/$A$6	
17				E7 = C7*($A$6+B7)/$A$6	
18					
19	B21:D24 = LINEST(E7:E15,D7:D15,TRUE,TRUE)				
20					
21		LINEST output:			
22		m 0.646299952	1.8687043	b	
23		s_m 0.009980942	0.0374181	s_b	
24		R^2 0.998333334	0.0644701	s_y	
25					
26	x-intercept =	-2.89139	= -b/m		
27					
28	n =	9	= COUNT(B7:B15)		
29	Mean y =	3.852	= AVERAGE(E7:E15)		
30	$\Sigma(x_i - \text{mean } x)^2$ =	41.722776	= DEVSQ(D7:D15)		
31					
32	Std deviation of				
33	x-intercept =	0.09787			
34	B33 = (C24/ABS(B22))*SQRT((1/B28) + B29^2/(B22^2*B30))				

17-19. Moles of ascorbic acid in experiment = (50 mL)(2.4 mM) = 0.12 mmol

If 3.9 µA flows for 10 min, mol e⁻ = It/F = (3.9 × 10⁻⁶ C/s)(600 s)/(96 485 C/mol)

= 2.4×10^{-8} mol e⁻. Two electrons come from oxidation of 1 molecule of ascorbic acid, so only 1.2×10^{-8} mol ascorbic acid, or 0.010% of the total ascorbic acid present, is consumed. The concentration is essentially constant.

17-20. The dropping-mercury electrode gives a fresh, reproducible surface for each measurement. Reduction of H⁺ is difficult on a Hg surface and occurs at much more negative potentials than it does on other electrodes. Therefore, Hg has a larger potential range than do most other electrodes for studying reductions. By contrast, Hg is easily oxidized to Hg^{2+} at modest positive potentials, so it is not useful for studying oxidation reactions that require a potential more positive than ~0.1 V vs S.C.E.

17-21. The spreadsheet shows that the least squares parameters for a graph of I_d vs $[Cu^{2+}]$ are

slope = m = 6.616   standard deviation = 0.018
intercept = b = −0.086   standard deviation = 0.062
s_y = 0.142

An unknown giving a current of 15.6 μA has a concentration of

$$[Cu^{2+}] = \frac{(I_d - b)}{m} = \frac{15.6 - (-0.086)}{6.616} = 2.371 \text{ mM}$$

uncertainty = ±0.023 ⇒ $[Cu^{2+}]$ = 2.37 (±0.02) mM (cells B24 and B25)

	A	B	C	D
1	Least-Squares Spreadsheet			
2		x =		
3	Highlight cells B14:C16	$[Cu^{2+}]$	y	
4	Type "= LINEST(C4:C11,	0.0393	0.256	
5	B4:B11,TRUE,TRUE)	0.078	0.520	
6	For PC, press	0.1585	1.058	
7	CTRL+SHIFT+ENTER	0.489	3.06	
8	For Mac, press	0.990	6.37	
9	COMMAND+RETURN	1.97	13.00	
10		3.83	25.0	
11		8.43	55.8	
12				
13		LINEST output:		
14		m	6.6157	-0.0858 b
15		s_m	0.0185	0.0622 s_b
16		R^2	1.0000	0.1416 s_y
17				
18		n =	8	B18 = COUNT(B4:B11)
19		Mean y =	13.133	B19 = AVERAGE(C4:C11)
20		$\Sigma(x_i - \text{mean } x)^2$ =	58.93	B20 = DEVSQ(B4:B11)
21				
22		Measured y =	15.6	Input
23		k = Number of replicate measurements of y =	1	Input
24		Derived x =	2.371001	B24 = (B22-C14)/B14
25		s_x =	0.0227	B25 = (C16/B14)*SQRT((1/B23)+(1/B18)+((B22-B19)^2)/(B14^2*B20))

Graph: y = 6.6157x − 0.0858, with y = I_d plotted against x = log$[Cu^{2+}]$.

17-22. $\dfrac{[X]_i}{[S]_f + [X]_f} = \dfrac{I_X}{I_{S+X}}$

$$\dfrac{x \text{ (mM)}}{3.00\left(\dfrac{2.00}{52.00}\right) + x\left(\dfrac{50.0}{52.0}\right)} = \dfrac{0.37 \text{ μA}}{0.80 \text{ μA}} \Rightarrow x = 0.096 \text{ mM}$$

17-23. The relative heights of the signals for acetone are $\dfrac{[X]_i}{[S]_f + [X]_f} = \dfrac{I_X}{I_{S+X}} = 0.25_9$.

Both samples contain the same concentration of methanol, so $[X]_i = [X]_f + x$.

$\dfrac{x \text{ (wt\%)}}{0.001\,00 + x} = 0.25_9 \Rightarrow x = 0.000\,35$ wt% acetone

17-24. (a) Concentration (deposition) stage: $Cu^{2+} + 2e^- \rightarrow Cu(s)$

(b) Stripping stage: $Cu(s) \rightarrow Cu^{2+} + 2e^-$

17-25. The spreadsheet gives a concentration of 0.227 ppb (cell B18) with a standard deviation of 0.039 ppb (cell B24) for the 20.0 mL-solution. The blood concentration is (20.0 mL/0.50 mL) = 40 times higher = 40(0.227 ± 0.039 ppb) = 9.1 ± 1.6 ppb.

	A	B	C	D
1	Standard Addition Constant Volume Least-Squares Spreadsheet			
2				
3				
4	x	y		
5	Added Cr			
6	(ppb)	Signal		
7	0.00	2.9		
8	0.26	7.0		
9	0.66	12.0		
10				
11	B14:C16 = LINEST(B7:B9,A7:A9,TRUE,TRUE)			
12				
13		LINEST output:		
14	m	13.6791	3.1051	b
15	s_m	0.8880	0.3637	s_b
16	R^2	0.9958	0.4175	s_y
17				
18	x-intercept = -b/m =	-0.227		
19				
20	n =	3	B20 = COUNT(A7:A9)	
21	Mean y =	7.300	B21 = AVERAGE(B7:B9)	
22	$\Sigma(x_i - \text{mean } x)^2 =$	0.221067	B22 = DEVSQ(A7:A9)	
23	Std deviation of			
24	x-intercept =	0.039		
25	B24 =(C16/ABS(B14))*SQRT((1/B20) + B21^2/(B14^2*B22))			

Plot: y = 13.68x + 3.11, Intercept = -0.227 ppb, axes "Added Sr (ng/mL)" vs "Signal".

17-26. (a) $H_2SO_3 \underset{}{\overset{pK_1 = 1.86}{\rightleftharpoons}} HSO_3^- \underset{}{\overset{pK_2 = 7.17}{\rightleftharpoons}} SO_3^{2-}$

H_2SO_3 is predominant below pH 1.86. HSO_3^- dominates between pH 1.86 and 7.17. SO_3^{2-} is dominant above pH 7.17.

(b) Cathode: $H_2O + e^- \rightarrow \frac{1}{2}H_2(g) + OH^-$

Anode: $3I^- \rightarrow I_3^- + 2e^-$

(c) $I_3^- + HSO_3^- + H_2O \rightarrow 3I^- + SO_4^{2-} + 3H^+$

$$I_3^- + 2S_2O_3^{2-} \rightleftharpoons 3I^- + \underset{\text{Tetrathionate}}{{}^-O\overset{O}{\underset{\|}{\overset{\|}{S}}}-S-S-\overset{O}{\underset{\|}{\overset{\|}{S}}}O^-}$$

Thiosulfate   Tetrathionate

(d) In step 3, I_3^- was generated by a current of 10.0 mA (= 10.0×10^{-3} C/s) for 4.00 min (= 240 s).

charge = $I \cdot t = (10.0 \times 10^{-3}$ C/s$)(240$ s$) = 2.40$ C

mol $e^- = I/F = (2.40$ C$)/(96\,485$ C/mol$) = 24.8_7$ μmol e^-

The anode reaction generates $\frac{1}{2}$ mol I_3^- for 1 mol e^-. Therefore, 24.8_7 μmol e^- will generate $\frac{1}{2}(24.8_7) = 12.4_4$ μmol I_3^-.

In step 5, 0.500 mL of 0.050 7 M thiosulfate = 25.3_5 μmol $S_2O_3^{2-}$. But 2 mol $S_2O_3^{2-}$ consume 1 mol I_3^-. Therefore, 25.3_5 μmol $S_2O_3^{2-}$ will consume $\frac{1}{2}(25.3_5) = 12.6_8$ μmol I_3^-.

We added excess $S_2O_3^{2-}$ in step 5 and consumed the excess in step 6. In step 6, we had to generate I_3^- at 10.0 mA for 131 s to react with excess $S_2O_3^{2-}$.

charge = $I \cdot t = (10.0 \times 10^{-3}$ C/s$)(131$ s$) = 1.31$ C

mol $e^- = I/F = (1.31$ C$)/(96\,485$ C/mol$) = 13.5_8$ μmol e^-

mol $I_3^- = \frac{1}{2}(13.5_8) = 6.79$ μmol I_3^-

Here is where we are so far:

Step 3: 12.4_4 μmol I_3^- were generated.

Step 4: x μmol I_3^- were consumed by sulfite in wine.

Step 5: We added enough $S_2O_3^{2-}$ to consume 12.6_8 μmol I_3^-.

Step 6: We had to generate 6.79 μmol I_3^- to consume excess $S_2O_3^{2-}$ from step 5. Therefore, I_3^- left after step 4 = $12.6_8 - 6.79 = 5.8_9$ μmol I_3^-

We began with 12.4_4 μmol I_3^-, and 5.89 μmol I_3^- were left after reaction with sulfite in wine. Therefore, sulfite in wine consumed $12.4_4 - 5.8_9 = 6.5_5$ µmol I_3^-. But 1 mol I_3^- reacts with 1 mol sulfite. Therefore, the wine contained 6.5_5 µmol sulfite in the 2.00 mL injected for analysis.

The wine sample prepared in step 1 consisted of 9.00 mL wine diluted to 10.00 mL. Therefore the original wine contained 10.00/9.00 of the amount found in the analysis. That is, 2.000 mL of pure wine contains $(10.00/9.00)(6.5_5 \text{ µmol sulfite}) = 7.2_8$ µmol sulfite.

$$\text{sulfite in wine} = \frac{7.2_8 \text{ µmol sulfite}}{2.00 \text{ mL}} = 3.64 \text{ mM}$$

This problem left out a description of the blank titration that should be done in a real analysis. There are components in wine in addition to sulfite that could react with I_3^-. For the blank titration, 1 M formaldehyde is added to the wine to bind all sulfite. The sulfite-formaldehyde adduct is not decomposed in 2 M NaOH and does not react with I_3^-. The blank titration consists of taking this formaldehyde/wine solution through the entire procedure. We subtract I_3^- consumed by the blank from I_3^- consumed by the wine without formaldehyde.

17-27. (a) Because all nitrite is converted to NO, we can write a proportionality in terms of grams instead of moles per liter:

$$\frac{\text{µg NO}_2^- \text{ in unknown}}{\text{µg NO}_2^- \text{ in standard} + \text{µg NO}_2^- \text{ in unknown}} = \frac{8.9 \text{ µA}}{14.6 \text{ µA}}$$

$$\frac{x}{5.00 + x} = \frac{8.9 \text{ µA}}{14.6 \text{ µA}} \Rightarrow \text{µg NO}_2^- \text{ in unknown} = 7.8_1$$

This nitrite was found in a 5.00-mL aliquot, so there were $\left(\frac{200.0}{5.00}\right)(7.8_1 \text{ µg})$

= 312 µg in 10.0 g of bacon = 31.2 µg of nitrite per gram of bacon.

(b) In both experiments, the nitrate increased the current in step 6 by 14.3 µA over that in step 5. From the first experiment, this response means that

$$\frac{\text{moles of NO}_3^- \text{ in unknown}}{\text{moles of NO}_2^- \text{ in unknown}} = \frac{14.3 \text{ µA}}{8.9 \text{ µA}} = 1.60_7$$

The formula mass of NO_2^- is 46.00, whereas that of NO_3^- is 62.00.

Because the bacon contains 31.2 µg of nitrate per gram, it must contain

$$\left(\frac{62.00 \text{ g NO}_3^-/\text{mol NO}_3^-}{46.00 \text{ g NO}_2^-/\text{mol NO}_2^-}\right)\left(1.60_7 \frac{\text{mol NO}_3^-}{\text{mol NO}_2^-}\right)(31.2 \text{ µg NO}_2^-)$$

= 67.6 µg nitrate per gram of bacon

CHAPTER 18
LET THERE BE LIGHT

18-1. (a) double (b) halve (c) double

18-2. (a) $E = h\nu = hc/\lambda = (6.626\,2 \times 10^{-34}\text{ J s})(2.997\,9 \times 10^8\text{ m s}^{-1})/(650 \times 10^{-9}\text{ m})$
$= 3.06 \times 10^{-19}$ J/photon
$(3.06 \times 10^{-19}\text{ J/photon})(6.022 \times 10^{23}\text{ photons/mol}) = 1.84 \times 10^5$ J/mol
$\left(1.84 \times 10^5 \dfrac{\text{J}}{\text{mol}}\right)\left(\dfrac{1\text{ kJ}}{1\,000\text{ J}}\right) = 184 \dfrac{\text{kJ}}{\text{mol}}$

(b) For $\lambda = 400$ nm, $E = 4.97 \times 10^{-19}$ J/photon = 299 kJ/mol.

18-3. $1\text{ eV} = 1.602 \times 10^{-19}$ J; $15\text{ keV} = (15 \times 10^3\text{ eV})\left(\dfrac{1.602 \times 10^{-19}\text{ J}}{1\text{ eV}}\right) = 2.403 \times 10^{-15}$ J

$E = h\nu = h\dfrac{c}{\lambda} \Rightarrow \lambda = \dfrac{hc}{E} = \dfrac{(6.626 \times 10^{-34}\text{ J·s})(2.998 \times 10^8\text{ m/s})}{2.403 \times 10^{-15}\text{ J}} = 8.266 \times 10^{-11}$ m

$\lambda\text{ (in nm)} = (8.266 \times 10^{-11}\text{ m})\left(\dfrac{10^9\text{ nm}}{\text{m}}\right) = 0.082\,66$ nm

18-4.

	Wavelength	Color absorbed	Color transmitted
(a)	450 nm	blue	orange
(b)	550 nm	yellow or yellow-green	violet-blue or violet
(c)	650 nm	red	blue-green

18-5. Transmittance (T) is the fraction of incident light that is transmitted by a substance: $T = P/P_0$, where P_0 is incident power and P is transmitted power. Absorbance is logarithmically related to transmittance: $A = -\log T$. When all light is transmitted, absorbance is zero. When no light is transmitted, absorbance is infinite. Absorbance is proportional to concentration. Molar absorptivity is the constant of proportionality between absorbance and the product cb, where c is concentration and b is pathlength.

18-6. Absorbance or molar absorptivity versus wavelength

18-7. The color of transmitted light is the complement of the color that is absorbed. If blue-green light is absorbed, the eye perceives the transmitted red light.

18-8.

Solution	Absorption maximum (nm)	Color expected from absorption maximum	Observed color
A	410	green-yellow	yellow
B	470	orange or red	orange
C	570	violet-blue	violet
D	710	green	blue

Solution D is predicted to be green, but the observed color is blue. According to the table, a blue solution should absorb orange light in range 580–620 nm and a blue-green solution should absorb red light in the range 620–680 nm. Solution D has substantial absorption in the 580–620 and 620–680 nm regions. The observed result of the broad absorption is a blue appearance.

18-9. Using $\nu = c/\lambda$, $\tilde{\nu} = 1/\lambda$ and $E = h\nu$, we find

	250 nm	2.5 μm
ν (Hz)	1.20×10^{15}	1.20×10^{14}
$\tilde{\nu}$ (cm^{-1})	4.00×10^{4}	4 000
J/photon	7.95×10^{-19}	7.95×10^{-20}
kJ/mol	479	47.9

18-10. $E = h\nu = hc/\lambda = (6.626\ 2 \times 10^{-34}\ \text{J s})(2.997\ 9 \times 10^{8}\ \text{m s}^{-1})/(10.6 \times 10^{-6}\ \text{m})$
$= 1.874 \times 10^{-20}$ J/photon

$5.0\ \text{kW} = 5.0\ \text{kJ/s} \Rightarrow \dfrac{5.0 \times 10^{3}\ \text{J/s}}{1.874 \times 10^{-20}\ \text{J/photon}} = 2.7 \times 10^{23}$ photons/s

(almost half a mole of photons per second)

18-11. $A = -\log T$. For example, if $T = 0.99$, $A = -\log(0.99) = 0.004\ 4$.

Transmittance:	0.99	0.90	0.50	0.10	0.010	0.001 0	0.000 10
Absorbance:	0.004 4	0.046	0.30	1.00	2.00	3.00	4.00

18-12. For a molar absorptivity (ε) of 1.00×10^{2} M^{-1} cm^{-1},
$A = \varepsilon bc = (100\ \text{M}^{-1}\ \text{cm}^{-1})(2.00\ \text{cm})(0.002\ 40\ \text{M}) = 0.480$.

For a molar absorptivity (ε) of 2.00×10^{2} M^{-1} cm^{-1},
$A = (200\ \text{M}^{-1}\ \text{cm}^{-1})(2.00\ \text{cm})(0.002\ 40\ \text{M}) = 0.960$.

Let There Be Light   173

Transmittance is obtained by raising 10 to the power on each side of the equation:

$$\log T = -A$$

$$10^{\log T} = 10^{-A}$$

$10^{\log T}$ is the same as T

For $A = 0.480$, $T = 10^{-0.480} = 0.331 = 33.1\%$
For $A = 0.960$, $T = 10^{-0.960} = 0.110 = 11.0\%$

Transmittance is not linearly related to absorbance or concentration. Doubling the absorbance does not decrease transmittance by a factor of 2.

18-13. SPF = $1/T$. Therefore, $T = 1/\text{SPF} = \frac{1}{2} = 0.5 = 50\%$. Half of the UV-B radiation is transmitted through the sunscreen. $A = -\log T = -\log 0.5 = 0.30$.
For SPF = 10, $T = 1/10 = 10\%$. $A = -\log T = -\log(1/10) = 1$.
For SPF = 20, $T = 1/20 = 5\%$. $A = -\log T = -\log(1/20) = 1.30$.
The SPF 10 sunscreen absorbs 90%, and SPF 20 absorbs 95% of UB-B radiation.

18-14. $\varepsilon = A/bc = 0.822/[(1.00 \text{ cm})(2.31 \times 10^{-5} \text{ M})] = 3.56 \times 10^4 \text{ M}^{-1} \text{ cm}^{-1}$

18-15. (a) $c = A/\varepsilon b = 0.244/[(8.83 \times 10^4 \text{ M}^{-1} \text{ cm}^{-1})(0.100 \text{ cm})] = 2.76 \times 10^{-5} \text{ M}$

(b) $(2.76 \times 10^{-5} \text{ mol/L})(81\,000 \text{ g/mol}) = 2.24 \text{ g/L}$

18-16. (a) Concentration = $\dfrac{(15.0 \times 10^{-3} \text{ g})/(384.63 \text{ g/mol})}{5 \times 10^{-3} \text{ L}} = 7.80 \times 10^{-3} \text{ M}$

(b) One-tenth dilution $\Rightarrow 7.80 \times 10^{-4} \text{ M}$

(c) $\varepsilon = A/bc = 0.634/[(0.500 \text{ cm})(7.80 \times 10^{-4} \text{ M})] = 1.63 \times 10^3 \text{ M}^{-1} \text{ cm}^{-1}$

18-17. (a) Absorbance measured on the graph at 350 nm is 0.052.

(b) $T = 10^{-A} = 10^{-0.052} = 0.89$. 89% of the light passes through the sunscreen.

18-18. (a) $A = -\log(P/P_0) = -\log T = -\log(0.45) = 0.347$

(b) The absorbance will double to 0.694, thereby giving $T = 10^{-A} = 10^{-0.694} = 0.202 \Rightarrow \%T = 20.2\%$.

18-19. Original concentration = $\dfrac{(0.267 \text{ g})/(337.69 \text{ g/mol})}{(0.100\ 0 \text{ L})} = 7.91 \times 10^{-3}$ M

Diluted concentration = $\left(\dfrac{2.000 \text{ mL}}{100.0 \text{ mL}}\right)(7.91 \times 10^{-3} \text{ M}) = 1.58 \times 10^{-4}$ M

$\varepsilon = \dfrac{A}{bc} = \dfrac{0.728}{(1.58 \times 10^{-4} \text{ M})(2.00 \text{ cm})} = 2.30 \times 10^{3}$ M^{-1} cm^{-1}

18-20. (a) Relation between absorbance and transmittance: $A = -\log T$
24.4% T corresponds to $A = -\log 0.244 = 0.612_6$

(b) Ideal gas law: $PV = nRT$; Molar concentration = $\dfrac{\text{moles}}{\text{volume}} = \dfrac{n}{V} = \dfrac{P}{RT}$

To convert 30.3 μbar = 30.3×10^{-6} bar to mol/L, we write

$\dfrac{n}{V} = \dfrac{P}{RT} = \dfrac{30.3 \times 10^{-6} \text{ bar}}{\left(0.083\ 145 \dfrac{\text{L·bar}}{\text{mol·K}}\right)(298 \text{ K})} = 1.22_3 \times 10^{-6}$ M

(c) Molar absorptivity = $\varepsilon = \dfrac{A}{bc}$. The pathlength is $b = 3.00$ cm and the concentration is $c = 1.22_3 \times 10^{-6}$ M.

$\varepsilon = \dfrac{A}{bc} = \dfrac{0.612_6}{(3.00 \text{ cm})(1.22_3 \times 10^{-6} \text{ M})} = 1.67 \times 10^{5}$ M^{-1} cm^{-1}

18-21. (a) $\varepsilon = \dfrac{A}{cb} = \dfrac{0.624 - 0.029}{(3.96 \times 10^{-4} \text{ M})(1.000 \text{ cm})} = 1.50 \times 10^{3}$ M^{-1} cm^{-1}

(b) $c = \dfrac{A}{\varepsilon b} = \dfrac{0.375 - 0.029}{(1.50 \times 10^{3} \text{ M}^{-1} \text{ cm}^{-1})(1.000 \text{ cm})} = 2.31 \times 10^{-4}$ M

(c) $c = \dfrac{A}{\varepsilon b} = \dfrac{0.733 - 0.029}{(1.50 \times 10^{3} \text{ M}^{-1} \text{ cm}^{-1})(1.000 \text{ cm})} = 4.69 \times 10^{-4}$ M = M_{dil}

(d) $M_{dil} V_{dil} = M_{conc} V_{conc}$

$\Rightarrow M_{conc} = M_{dil} \dfrac{V_{dil}}{V_{conc}} = (4.69 \times 10^{-4} \text{ M})\left(\dfrac{25.00 \text{ mL}}{2.00 \text{ mL}}\right) = 5.87 \times 10^{-3}$ M

18-22. (a) $c = A/\varepsilon b = 0.427/[(6\ 130 \text{ M}^{-1} \text{ cm}^{-1})(1.000 \text{ cm})] = 6.97 \times 10^{-5}$ M

(b) The sample had been diluted × 10 $\Rightarrow 6.97 \times 10^{-4}$ M

(c) $\dfrac{x \text{ g}}{(292.16 \text{ g/mol})(5 \times 10^{-3} \text{ L})} = 6.97 \times 10^{-4}$ M $\Rightarrow x = 1.02$ mg

Let There Be Light

18-23. $[Fe]_{\text{in reference cell}} = \left(\dfrac{10.0}{50.0}\right)(6.80 \times 10^{-4}\text{ M}) = 1.36 \times 10^{-4}\text{ M}$

Setting the absorbances of sample and reference equal to each other in Beer's law gives $\varepsilon_s b_s c_s = \varepsilon_r b_r c_r$.

But $\varepsilon_s = \varepsilon_r$, so $b_s c_s = b_r c_r \Rightarrow (2.48\text{ cm})c_s = (1.00\text{ cm})(1.36 \times 10^{-4}\text{ M})$
$\Rightarrow c_s = 5.48 \times 10^{-5}\text{ M}.$

This is a 1/4 dilution of runoff, so $[Fe]_{\text{in runoff}} = 2.19 \times 10^{-4}\text{ M}.$

18-24.

Sample	Corrected absorbance	
0.538 ppm	0.098	
1.076 ppm	0.196	
2.152 ppm	0.390	
3.228 ppm	0.577	
4.034 ppm	0.732	
Unknown	0.310	average = 0.313_7
Unknown	0.316	standard deviation = 0.003_2
Unknown	0.315	

(a) ppm nitrite nitrogen (in cell B24 of the spreadsheet) = 1.73_5 ppm

uncertainty (in cell B25 of the spreadsheet) = $\pm 0.01_7$ ppm

(b) $\dfrac{1.735 \times 10^{-3}\text{ g N/L}}{14.006\ 7\text{ g N/mol}} = 1.239 \times 10^{-4}$ M nitrogen. The molar concentration of NO_2^- is the same as that of nitrogen = $1.23_9 \times 10^{-4}$ M.

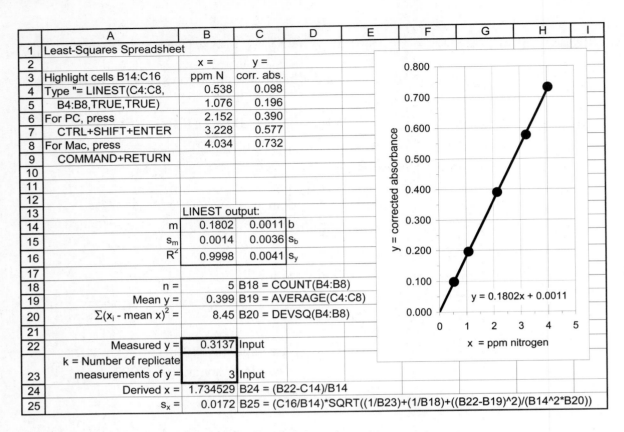

18-25. The ppm concentration of N in 0.015 83 M NaNO$_2$ is

$$\frac{\text{g N}}{\text{mL}} = \left(0.015\ 83\ \frac{\text{mol}}{\text{L}}\right)\left(14.007\ \frac{\text{g}}{\text{mol}}\right)\left(0.001\ \frac{\text{L}}{\text{mL}}\right) = 2.217 \times 10^{-4}\ \frac{\text{g}}{\text{mL}}$$

$$\text{ppm} = \left(2.217 \times 10^{-4}\ \frac{\text{g}}{\text{mL}}\right)\left(10^6\ \frac{\mu\text{g}}{\text{g}}\right) = 221.7\ \frac{\mu\text{g}}{\text{mL}}$$

A 1:100 dilution (10.00 mL diluted to 1.000 L) would be 2.217 ppm. A dilution of 5.00 mL to 1.000 L would be 1.109 ppm. A 50% dilution of the 1.109 ppm solution would be 0.554 ppm. A dilution of 15.00 mL (= 5.00 + 10.00 mL) to 1.000 L would be 3.326 ppm.

18-26. (a) $NO_3^- + NADH + H^+ \xrightarrow[\text{pH 7}]{\text{nitrate reductase}} NO_2^- + NAD^+ + H_2O$

[Structural reaction: sulfanilamide ($H_2N-SO_2-C_6H_4-NH_2$) + NO_2^- + N-(1-naphthyl)ethylenediamine $\xrightarrow{H^+}$ red azo product]

The first reaction is catalyzed by the enzyme nitrate reductase.

(b)

	A	B	C	D	E
1	Least-Squares Spreadsheet				
2		x =	y =		
3	Highlight cells B14:C16	ppm N	absorbance		
4	Type "= LINEST(C4:C11,	0.250	0.062		
5	B4:B11,TRUE,TRUE)	0.500	0.069		
6	For PC, press	1.00	0.108		
7	CTRL+SHIFT+ENTER	1.50	0.126		
8	For Mac, press	2.50	0.209		
9	COMMAND+RETURN	5.00	0.423		
10		7.50	0.592		
11		10.00	0.761		
12					
13		LINEST output:			
14	m	0.0738	0.0335	b	
15	s_m	0.0014	0.0068	s_b	
16	R^2	0.9979	0.0132	s_y	
17					
18	n =	8	B18 = COUNT(B4:B11)		
19	Mean y =	0.294	B19 = AVERAGE(C4:C11)		
20	$\Sigma(x_i - \text{mean } x)^2 =$	91.30	B20 = DEVSQ(B4:B11)		
21					
22	Measured y =	0.1965	Input		
23	k = Number of replicate measurements of y =	2	Input		
24	Derived x =	2.210616	B24 = (B22-C14)/B14		
25	s_x =	0.1441			
26	B25 = (C16/B14)*SQRT((1/B23)+(1/B18)+((B22-B19)^2)/(B14^2*B20))				

From the average absorbance of 2 unknowns in cell B22, the concentration of nitrate nitrogen in the aquarium water is 2.21 ppm in cell B24 with an uncertainty of 0.14 ppm in cell B25.

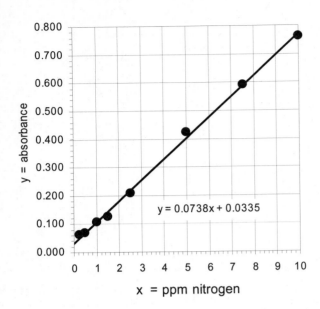

(c) Reactions in part (a) convert one atom of nitrate nitrogen into one molecule of colored product. Therefore the molarity of colored product in the final solution is the same as the molarity of nitrate N in the final solution. 1 ppm nitrate N means 1 μg nitrate N per gram of aqueous solution. If the density of aqueous solution is near 1.00 g/mL, a 1 ppm solution contains close to 1 μg N/mL = 1 mg N/L. The molarity of nitrate N in a 1 ppm solution is $(10^{-3}$ g N/14.006 7 g/mol)/L = 7.14×10^{-5} M nitrate N. Each 50 μL standard was diluted to 2.02 mL for analysis, so the concentration of nitrate N in the final solution whose absorbance was measured is approximately $(7.14 \times 10^{-5}$ M)(0.050 mL/2.02 mL) = 1.77×10^{-6} M. The slope of the calibration curve is 0.073 8 (ppm N)$^{-1}$. That is, the absorbance from a solution containing 1 ppm N is 0.073 8. The molar absorptivity of colored product is $\varepsilon = A/bc =$ 0.073 8/[(1.00 cm)(1.77 $\times 10^{-6}$ M)] = 4.2×10^{4} M^{-1}cm^{-1}.

18-27. The wt% of N in a solution containing 28.6 wt% NH$_3$ is

$$\text{wt\% NH}_3 = \left(\frac{14.006\ 7\ \text{g N/mol}}{17.030\ 5\ \text{g NH}_3/\text{mol}}\right)(28.6) = 23.5_2\%.$$

This solution contains 0.235_2 g of N per gram of solution.

1.000 g of N is contained in $\dfrac{1\ \text{g N}}{0.235_2\ \text{g N/g solution}} = 4.25_1$ g solution.

If 4.25_1 g of 28.6 wt% NH$_3$ is dissolved in 1.000 L, the solution will contain 1.000 g N/L = 1.000 mg N/mL = 1 000 ppm N.

To prepare 1.00 ppm N, dilute 1.000 mL to 1.000 L.

Let There Be Light

To prepare 2.00 ppm N, dilute 2.000 mL to 1.000 L.

To prepare 4.00 ppm N, dilute 4.000 mL to 1.000 L.

To prepare 8.00 ppm N, dilute 8.000 mL to 1.000 L.

18-28. (a) 1.00×10^{-2} g of NH_4Cl in 1.00 L = 1.869×10^{-4} M. To make the colored solution in step 2, 10.0 mL is diluted to 50.0 mL:

$M_{conc}V_{conc} = M_{dil}V_{dil}$

$(1.869 \times 10^{-4} \text{ M})(10.0 \text{ mL}) = M_{dil}(50.0 \text{ mL}) \Rightarrow M_{dil} = 3.739 \times 10^{-5}$ M

$\varepsilon = \dfrac{A}{bc} = \dfrac{0.308 - 0.140}{(1.00 \text{ cm})(3.739 \times 10^{-5} \text{ M})} = 4.49_3 \times 10^3$ M^{-1} cm^{-1}

(b) In step 2, the corrected absorbance is $0.592 - 0.140 = 0.452$.

$c = A/\varepsilon b = 0.452/[(4.49_3 \times 10^3 \text{ M}^{-1} \text{ cm}^{-1})(1.00 \text{ cm})] = 1.00_6 \times 10^{-4}$ M

(c) $M_{conc}V_{conc} = M_{dil}V_{dil}$

$M_{conc}(10.0 \text{ mL}) = (1.00_6 \times 10^{-4})(50.0 \text{ mL}) \Rightarrow M_{dil} = 5.03_0 \times 10^{-4}$ M

(d) 100.00 mL of unknown = 5.030×10^{-5} mol of N = 7.045×10^{-4} g of N

$\Rightarrow$ wt% N = $(7.044 \times 10^{-4} \text{ g})/(4.37 \times 10^{-3} \text{ g}) = 16.1$ wt%

18-29. (a) mg Cu in flask C = $(1.00 \text{ mg}) \times \left(\dfrac{10}{250}\right) \times \left(\dfrac{15}{30}\right) = 0.020\ 0$ mg

  Original   Fraction   Fraction
  sample   taken from taken from
  size   flask A   flask B

This entire quantity is in the isoamyl alcohol (20.00 mL), so the concentration is $[(2.00 \times 10^{-5} \text{ g})/(63.546 \text{ g/mol})]/(0.020\ 0 \text{ L}) = 1.57 \times 10^{-5}$ M.

(b) Observed absorbance = absorbance due to Cu in rock + blank absorbance = $\varepsilon bc + 0.056 = (7.90 \times 10^3)(1.00)(1.574 \times 10^{-5}) + 0.056 = 0.180$. The observed absorbance is equal to the absorbance from Cu in the rock *plus* the blank absorbance. In the lab, we measure the observed absorbance and subtract the blank absorbance from it to find the absorbance due to copper.

(c) $\dfrac{\text{Cu in unknown}}{\text{Cu in known}} = \dfrac{\text{absorbance of unknown}}{\text{absorbance of known}}$

$\dfrac{x \text{ mg}}{1.00 \text{ mg}} = \dfrac{0.874 - 0.056}{0.180 - 0.056} \Rightarrow x = 6.60$ mg Cu

18-30. (a) The concentration of phosphate in solution A is $1.19_6 \times 10^{-3}$ M. When 0.140 mL of A is diluted to 5.00 mL,

[phosphate] = $(1.19_6 \times 10^{-3} \text{ M})\left(\dfrac{0.140 \text{ mL}}{5.00 \text{ mL}}\right) = 3.34_9 \times 10^{-5}$ M

$\varepsilon = A/bc = (0.829 - 0.017)/[(1.00 \text{ cm})(3.34_9 \times 10^{-5} \text{ M})]$
$= 2.42_5 \times 10^4 \text{ M}^{-1} \text{ cm}^{-1}$

(b) [phosphate] in analyte $= \dfrac{A}{\varepsilon b} = \dfrac{(0.836 - 0.038)}{(2.42_5 \times 10^4 \text{ M}^{-1} \text{ cm}^{-1})(1.00 \text{ cm})}$

$= 3.29 \times 10^{-5}$ M

But this phosphate had been diluted from 0.300 mL to 5.00 mL. Therefore the original concentration of phosphate was

$M_{conc}V_{conc} = M_{dil}V_{dil}$

$M_{conc}(0.300 \text{ mL}) = (3.29 \times 10^{-5} \text{ M})(5.00 \text{ mL}) \Rightarrow M_{conc} = 5.48_5 \times 10^{-4}$ M

The original volume of digested ferritin was 1.00 mL, which contains $(1.00 \times 10^{-3} \text{ L})(5.48_5 \times 10^{-4} \text{ M}) = 5.48_5 \times 10^{-7}$ mol phosphorus $= 1.69_9 \times 10^{-5}$ g P.

wt% phosphorus $= 100 \times \dfrac{1.69_9 \times 10^{-5} \text{ g phosphorus}}{1.35 \text{ mg ferritin}} = 1.26$ wt%

CHAPTER 19
SPECTROPHOTOMETRY: INSTRUMENTS AND APPLICATIONS

19-1. In a single-beam spectrophotometer, we first place a cuvet containing a blank reference solution into the single sample holder and set the detector readout to indicate 100% transmittance. This readout corresponds to the radiant power P_0 reaching the detector. Then we place the sample solution in the holder, and the radiant power reaching the detector is taken as P. Transmittance is the quotient P/P_0. In a double-beam spectrophotometer, the sample and reference cuvets are both in the instrument. A rotating beam chopper alternately directs light from the monochromator to the sample or reference. Radiant power reaching the detector from the reference is taken as P_0 and radiant power reaching the detector from the sample is taken as P. The single-beam instrument produces errors if the source intensity or detector response drifts between measurement of the blank and the sample. This error is absent in the double-beam instrument, which measures light passing through both cuvets several times per second.

19-2. Deuterium lamp

19-3. Advantage: increased ability to resolve closely spaced spectral peaks
Disadvantage: more noise because less light reaches detector, thereby producing a lower signal-to-noise ratio

19-4. (a) $n\lambda = d(\sin \theta + \sin \phi)$
$1 \cdot 600 \times 10^{-9}$ m $= d(\sin 40° + \sin(-30°)) \Rightarrow d = 4.20 \times 10^{-6}$ m
Lines/cm $= 1/(4.20 \times 10^{-4}$ cm$) = 2.38 \times 10^3$ lines/cm

(b) $\lambda = 1/(1\,000$ cm$^{-1}) = 10^{-3}$ cm $\Rightarrow d = 7.00 \times 10^{-3}$ cm $\Rightarrow$ 143 lines/cm

19-5. (a) $n\lambda = d(\sin \theta + \sin \phi)$
Inserting the values $n = -1$, $\lambda = 633$ nm $= 0.633$ μm, $d = 1.6$ μm, and $\theta = 0°$ allows us to solve for ϕ:
$(-1)(0.633$ μm$) = (1.6$ μm$)(\sin 0° + \sin \phi) \Rightarrow \phi = -23°$
For $n = +1$, $\phi = +23°$. For $n = +2$, $\phi = +52°$.

(b) For the third order diffracted beam, we find
$n\lambda = d(\sin \theta + \sin \phi)$
$(3)(0.633$ μm$) = (1.6$ μm$)(\sin 0° + \sin \phi) \Rightarrow \sin \phi = 1.19$
But the sine of an angle cannot exceed 1, so the diffraction condition cannot be satisfied for $n = 3$.

19-6. An electron ejected from the photocathode is accelerated by a positive voltage so it strikes the first dynode with high kinetic energy. Each electron from the cathode liberates many electrons from the first dynode. Each of these electrons, in turn, is accelerated by a positive voltage so it liberates many electrons from the second dynode. After several such stages, each electron liberated at the cathode results in $\sim 10^6$ electrons striking the anode at the end of the chain.

19-7. The photodiode array spectrophotometer measures the entire spectrum at once, because each detector element of the diode array measures a different, narrow band of wavelengths. The spectrum is recorded rapidly relative to that of a dispersive spectrophotometer, which measures one wavelength at a time. The disadvantage of the photodiode array instrument is that its resolution is not as great as that of the dispersive instrument, because there can only be as many different wavelengths resolved as there are detector elements in the array.

19-8. If the spectra of two compounds with a constant total concentration cross at any wavelength, all mixtures with the same total concentration will go through that same point, called an isosbestic point. The appearance of isosbestic points in a chemical reaction is good evidence that we are observing one main species being converted to one other major species.

19-9. $b(\varepsilon_X' \varepsilon_Y'' - \varepsilon_Y' \varepsilon_X'') = (1.000)[(16\ 440)(6\ 420) - (3\ 870)(3\ 990)] = 9.01_0 \times 10^7$

$[X] = \dfrac{1}{D}(A' \varepsilon_Y'' - A'' \varepsilon_Y') = \dfrac{(0.957)(6\ 420) - (0.559)(3\ 870)}{9.01_0 \times 10^7} = 4.42 \times 10^{-5}\ \text{M}$

$[Y] = \dfrac{1}{D}(A'' \varepsilon_X' - A' \varepsilon_X'') = \dfrac{(0.559)(16\ 440) - (0.957)(3\ 990)}{9.01_0 \times 10^7} = 5.96 \times 10^{-5}\ \text{M}$

19-10.

	A	B	C	D	E
1	Spreadsheet for spectrophotometric analysis of a mixture				
2					
3			Absorptivity (M^-1 cm^-1)		Absorbance of
4		Wavelength (nm)	Compound X	Compound Y	mixture
5	λ'	272	16440	3870	0.957
6	λ''	327	3990	6420	0.559
7					
8	Pathlength (b, cm) =		1.000		
9					
10	D =	9.010E+07		B10 = C8*(C5*D6 – D5*C6)	
11	[X] =	4.418E-05		B11 = (E5*D6 – E6*D5)/B10	
12	[Y] =	5.962E-05		B12 = (E6*C5 – E5*C6)/B10	

Spectrophotometry: Instruments and Applications

19-11.

	A	B	C	D	E
1	Spreadsheet for spectrophotometric analysis of a mixture				
2					
3			Absorptivity (M^-1 cm^-1)		Absorbance of
4		Wavelength	Permanganate	Dichromate	mixture
5	λ'	266	420	4100	0.766
6	λ''	320	1680	1580	0.422
7					
8	Pathlength (b, cm) =			1.000	
9					
10	D =	–6.224E+06		B10 = C8*(C5*D6 – D5*C6)	
11	[Permanganate] =	8.353E-05		B11 = (E5*D6 – E6*D5)/B10	
12	[Dichromate] =	1.783E-04		B12 = (E6*C5 – E5*C6)/B10	

19-12. (a) $c = \dfrac{A}{\varepsilon b} = \dfrac{0.463}{(4\,170)(1.00)} = 1.11_0 \times 10^{-4}$ M

$= 8.99$ g/L $= 8.99$ mg transferrin/mL

[Fe] $= 2.22_0 \times 10^{-4}$ M $= 0.012\,4$ g/L $= 12.4$ µg/mL

(b) The spreadsheet shows that [transferrin (TRF)] = 7.30×10^{-5} M and [desferrioxamine (DFO)] = 5.22×10^{-5} M. The fraction of iron in transferrin (which binds two Fe^{3+} ions) is $\dfrac{2[\text{TRF}]}{2[\text{TRF}] + [\text{DFO}]} = 73.7\%$.

	A	B	C	D	E
1	Spreadsheet for spectrophotometric analysis of a mixture				
2					
3			Absorptivity (M^-1 cm^-1)		Absorbance of
4		Wavelength	Transferrin	Desferrioxamine	mixture
5	λ'	428	3540	2730	0.401
6	λ''	470	4170	2290	0.424
7					
8	Pathlength (b, cm) =			1.000	
9					
10	D =	–3.278E+06		B10 = C8*(C5*D6 – D5*C6)	
11	[TRF] =	7.299E-05		B11 = (E5*D6 – E6*D5)/B10	
12	[DFO] =	5.224E-05		B12 = (E6*C5 – E5*C6)/B10	

19-13. (a) $A = 2\,080[\text{HIn}] + 14\,200[\text{In}^-]$

(b) $[\text{HIn}] = x$; $[\text{In}^-] = 1.84 \times 10^{-4} - x$

$A = 0.868 = 2\,080x + 14\,200(1.84 \times 10^{-4} - x) \Rightarrow x = 1.44 \times 10^{-4}$ M

$pK_{\text{HIn}} = \text{pH} - \log\dfrac{[\text{In}^-]}{[\text{HIn}]} = 6.23 - \log\dfrac{(1.84 \times 10^{-4}) - (1.44 \times 10^{-4})}{1.44 \times 10^{-4}}$

$= 6.79$

19-14.

	A	B	C	D	E
1	Spreadsheet for finding pKa from spectral data				
2				A – A(HIn)	
3	Absorbance of	pH	A	A(In-) – A	log(column D)
4	In- at high pH =	3.35	0.170	0.253	–0.597
5	0.818	3.65	0.287	0.529	–0.276
6	Absorbance of	3.94	0.411	0.995	–0.002
7	HIn at low pH =	4.30	0.562	2.172	0.337
8	0.006	4.64	0.670	4.486	0.652
9					
10	D4 = (C4 – $A$8)/($A$5 –C4)				

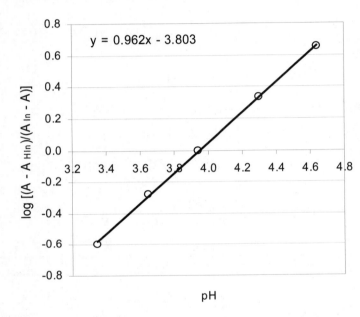

From the equation of the line, the x-intercept is

$$\text{pH} = \frac{3.803}{0.962} = 3.95.$$

Therefore, $pK_{\text{HIn}} = 3.95$.

19-15. First convert T to $A = -\log T$ and then convert A to $\varepsilon = \dfrac{A}{bc} = \dfrac{A}{(0.005\,00)(0.010\,0)}$

	Absorbance			ε (M^{-1} cm^{-1})	
	2 022	1 993 cm^{-1}		2 022	1 993 cm^{-1}
A	0.508 6	0.098 54	A	10 170	1 971
B	0.011 44	0.699 0	B	228.8	13 980

For the mixture, $A_{2022} = -\log(0.340) = 0.468\,5$ and $A_{1993} = -\log(0.383) = 0.416\,8$. Using Equations 19-5, we find [A] = 0.009 11 M and [B] = 0.004 68 M.

19-16. Theoretical equivalence point =

$$\frac{\left(2\,\dfrac{\text{mol Ga}}{\text{mol transferrin}}\right)\left(\dfrac{0.003\,57\text{ g transferrin}}{81\,000\text{ g transferrin/mol transferrin}}\right)}{0.006\,64\,\dfrac{\text{mol Ga}}{\text{L}}} = 13.3\ \mu\text{L}$$

Observed end point ≈ intersection of lines taken from first 6 points and last 4 points in the graph below = 12.2 μL, corresponding to $\dfrac{12.2}{13.3}$ = 91.7% of 2 Ga/transferrin = 1.83 Ga/transferrin. In the absence of oxalate, there is no evidence for specific binding of Ga to the protein, because the slope of the curve is small and does not change near 1 or 2 Ga/transferrin.

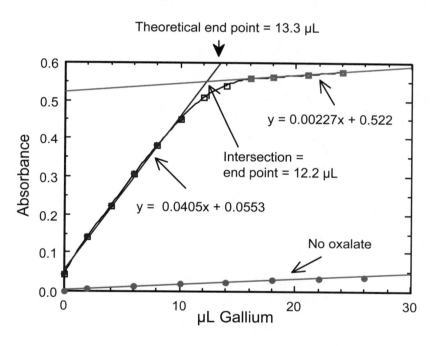

19-17. Absorbance is corrected by multiplying each observed absorbance by (total volume/initial volume). For example, at 36.0 μL, A(corrected) = (0.399)[(2 025 + 36)/2 025] = 0.406. A graph of corrected absorbance versus volume of Pb^{2+} (μL) is similar to Figure 19-16, with the end point at 46.7 μL. The moles of Pb^{2+} in this volume are
(46.7 × 10^{-6} L)(7.515 × 10^{-4} M) = 3.510 × 10^{-8} mol.
The concentration of semi-xylenol orange is
(3.510 × 10^{-8} mol)/(2.025 × 10^{-3} L) = 1.73 × 10^{-5} M.

19-18.

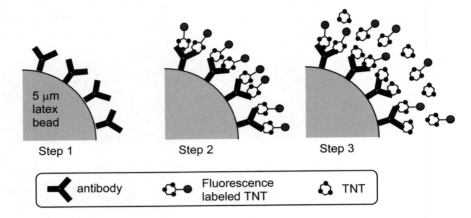

In step 1, antibodies for TNT are attached to latex beads. In step 2, the antibody is saturated with a fluorescent derivative of TNT. Excess fluorescent derivative is removed. In step 3, the beads are incubated with TNT, which displaces some fluorescent derivative from binding sites on the antibodies. The suspension of beads is then injected into the flow cytometer. As each bead passes in front of the detector, it is excited by a laser and its fluorescence is measured. The graph in the textbook shows median bead fluorescence versus TNT concentration in a series of standards. The more TNT in the standard, the less fluorescence remains associated with the beads.

19-19. (a) The equation for standard addition in Section 5-3 is

$$I_{S+X}\left(\frac{V}{V_0}\right) = I_X + \frac{I_X}{[X]_i}[S]_i\left(\frac{V_s}{V_0}\right)$$

$\underbrace{\phantom{I_{S+X}\left(\frac{V}{V_0}\right)}}_{\text{Function to plot on } y\text{-axis}}$ $\underbrace{\phantom{[S]_i\left(\frac{V_s}{V_0}\right)}}_{\text{Function to plot on } x\text{-axis}}$

where V_0 is the volume of unknown in the cuvet (2.00 mL), V_s is the volume of standard added (0 to 40 μL), V is the total volume of unknown plus added standard, $[S]_i$ is the initial concentration of standard (1.40 μg Se/mL), $[X]_i$ is the initial concentration of unknown in the 2.00-mL solution, I_X is the fluorescence intensity from the unknown, and I_{S+X} is the fluorescence intensity from the unknown plus standard addition. A graph of $I_{S+X} V/V_0$ versus $[S]_i V_s/V_0$ is given after the spreadsheet below.

	A	B	C	D	E
1	Standard addition of Se for fluorescence analysis				
2					
3	Add 1.40 mg/mL Se standard to 2.00 mL of unknown				
4					
5	V_o (mL) =	V_s =	x-axis function	$I(s+x)$ =	y-axis function
6	2.00	mL Se added	S_i*V_s/V_o	signal	$I_{s+x}*V/V_o$
7	$[S]_i$ (ppm) =	0.0000	0.00000	41.4	41.40
8	1.40	0.0100	0.00700	49.2	49.45
9		0.0200	0.01400	56.4	56.96
10		0.0300	0.02100	63.8	64.76
11		0.0400	0.02800	70.3	71.71
12	Highlight cells B16:C18				
13	B16:C18 = LINEST(E7:E11,C7:C11,TRUE,TRUE)				
14	CTRL+SHIFT+ENTER (PC) or COMMAND+RETURN (Mac)				
15		LINEST output:			
16	m	1084.7143	41.6700	b	
17	s_m	14.8834	0.2552	s_b	
18	R^2	0.9994	0.3295	s_y	
19					
20	x-intercept = -b/m =	-0.03842			
21					
22	n =	5	B22 = COUNT(C7:C11)		
23	Mean y =	56.85600	B23 = AVERAGE(E7:E11)		
24	$\Sigma(x_i - \text{mean } x)^2$ =	0.00049	B24 = DEVSQ(C7:C11)		
25					
26	Std deviation of				
27	x-intercept =	0.000732			
28	B27 =(C18/ABS(B16))*SQRT((1/B22) + B23^2/(B16^2*B24))				

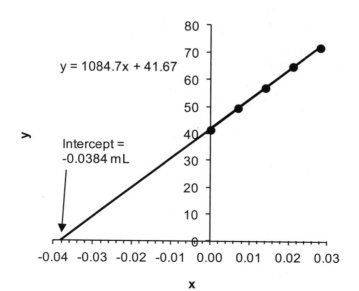

The x-intercept of the graph is computed from the equation of the straight line by setting $y = 0$: $0 = 1\,084.6x + 41.67 \Rightarrow x = -0.038\,4$. The concentration of Se in the unknown is 0.038 4 µg/mL. All of the selenium from 0.108 g of Brazil nuts was dissolved in 10.0 mL of solvent, which contained (10.0 mL)(0.038 4 µg/mL) = 0.384 µg Se. The weight percent of Se in the nuts is $100 \times (0.384 \times 10^{-6} \text{ g}/0.108 \text{ g}) = 3.56 \times 10^{-4}$ wt%.

(b) The standard deviation of the x-intercept is

$$\text{Standard deviation of intercept} = \frac{s_y}{|m|} \sqrt{\frac{1}{n} + \frac{\bar{y}^2}{m^2 \Sigma(x_i - \bar{x})^2}}$$

where s_y is the standard deviation of y, $|m|$ is the absolute value of the slope, n is the number of data points ($n = 5$), $\bar{y}$ is the mean value of y for the five points, x_i are the individual values of x for the five points, and $\bar{x}$ is the mean value of x for the five points. The standard deviation is evaluated in cell B27 of the spreadsheet in part (a).

The relative uncertainty in the intercept is $0.000\,732/0.038\,42 = 1.9\%$. This value would be the relative uncertainty in wt% Se if other sources of error are insignificant.

Uncertainty in wt% = $(0.019)(3.56 \times 10^{-4}$ wt%$) = 6.8 \times 10^{-6}$ wt%

Answer: $3.56\,(\pm 0.07) \times 10^{-4}$ wt%

19-20. We seek the quotient [In⁻]/[HIn], which is given by

$$\frac{[\text{In}^-]}{[\text{HIn}]} = \left(\frac{R - R_{\text{HIn}}}{R_{\text{In}^-} - R}\right) \frac{\alpha''_{\text{HIn}}}{\alpha''_{\text{In}^-}} = \left(\frac{1.080 - 3.14_5}{0.332 - 1.080}\right)(0.497_9) = 1.37$$

Putting this quotient into the Henderson-Hasselbalch equation for the acid, HIn, we find

$$\text{pH} = \text{p}K_a + \log\frac{[\text{In}^-]}{[\text{HIn}]} = 7.50 + \log(1.37) = 7.64$$

19-21. One molecule of antibody 2 in the drawing becomes attached to one molecule of analyte. Each molecule of enzyme bound to antibody 2 then makes many copies of a colored or fluorescent product that is detected. By this means, the signal available from one molecule of analyte is greatly amplified.

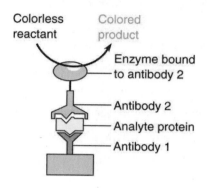

19-22. $\dfrac{570 \text{ kW·h}}{2 \text{ kW·h/kg coal}} = 285 \text{ kg coal}$ $\dfrac{0.6 \text{ kg C}}{\text{kg coal}} \times 285 \text{ kg coal} = 171 \text{ kg C}$

$\dfrac{44 \text{ kg CO}_2}{12 \text{ kg C}} \times 171 \text{ kg C} = 630 \text{ kg CO}_2$

$\dfrac{0.02 \text{ kg S}}{\text{kg coal}} \times 285 \text{ kg coal} = 5.7 \text{ kg S}$ $\dfrac{64 \text{ kg SO}_2}{32 \text{ kg S}} \times 5.7 \text{ kg S} = 11 \text{ kg SO}_2$

19-23. As VO^{2+} is added (traces 1–9), the peak at 439 decreases and a new one near 485 nm develops, with an isosbestic point at 457 nm. When VO^{2+}/xylenol orange > 1, the peak near 485 nm decreases and a new one at 566 nm grows in, with an isosbestic point at 528 nm. This sequence is logically interpreted by the sequence

$$\underset{}{M} + \underset{434 \text{ nm}}{L} \rightarrow \underset{485 \text{ nm}}{ML}$$

$$\underset{485 \text{ nm}}{ML} + \underset{}{M} \rightarrow \underset{566 \text{ nm}}{M_2L}$$

where M is vanadyl ion and L is xylenol orange. The structure of xylenol orange in Table 13-2 shows that it has metal-binding groups on both sides of the molecule and could form an M_2L complex.

CHAPTER 20
ATOMIC SPECTROSCOPY

20-1. In comparison with flames, furnaces give lower detection limits and require smaller sample volumes, but give poorer reproducibility with manual sample introduction. Automated sample introduction gives improved precision.

20-2. The inductively coupled plasma operates at much higher temperature than a conventional flame. This temperature lessens chemical interference (such as oxide formation) and allows emission instead of absorption to be used. Therefore, lamps are not required for each element. Self-absorption is reduced in the plasma because the temperature is more uniform. Disadvantages of the plasma are increased cost of equipment and operation.

20-3. Temperature is more critical in emission spectroscopy because the small population of the excited state varies substantially as the temperature is changed. The population of the ground state does not vary much.

20-4. In atomic emission, emission from analyte is superimposed on broad background emission. By measuring the emission at the peak and subtracting the emission slightly away from the peak, we can find the true analyte emission. In atomic absorption, we measure the absorbance of analyte + background at the peak wavelength of the hollow-cathode lamp. If we shift the monochromator away from the peak, there is virtually no light intensity coming from the hollow-cathode lamp and there will be nothing to measure.

20-5. A matrix modifier used with a graphite furnace can retard evaporation of analyte during charring or increase the evaporation of matrix during charring. The net effect is to reduce the interference of the matrix during atomization of analyte.

20-6. (a) $\Delta E = h\nu = \dfrac{hc}{\lambda} = \dfrac{(6.626\ 1 \times 10^{-34}\ \text{J·s})(2.997\ 9 \times 10^8\ \text{m/s})}{422.7 \times 10^{-9}\ \text{m}}$

$= 4.699 \times 10^{-19}\ \text{J}$

(b) $\dfrac{N^*}{N_0} = \dfrac{g^*}{g_0} e^{-\Delta E/kT} = 3e^{-(4.699 \times 10^{-19}\ \text{J})/(1.381 \times 10^{-23}\ \text{J/K})(2\ 500\ \text{K})}$

$= 3.67 \times 10^{-6}$

(c) At 2 515 K, $N^*/N_0 = 3.98 \times 10^{-6} \Rightarrow$ 8.4% increase from 2 500 to 2 515 K

(d) At 6 000 K, $N^*/N_0 = 1.03 \times 10^{-2}$

Atomic Spectroscopy

20-7. (a) $\Delta E = h\nu = \dfrac{hc}{\lambda} = \dfrac{(6.626\,1\times10^{-34}\,\text{J·s})(2.997\,9\times10^{8}\,\text{m/s})}{327\times10^{-9}\,\text{m}}$

$= 6.07 \times 10^{-19}$ J

(b) $\dfrac{N^*}{N_0} = \dfrac{g^*}{g_0} e^{-\Delta E/kT} = 3e^{-(6.07\,\times\,10^{-19}\,\text{J})/(1.381\,\times\,10^{-23}\,\text{J/K})(2\,400\text{K})}$

$= 3.2_9 \times 10^{-8}$

(c) At 2 415 K, $N^*/N_0 = 3.6_9 \times 10^{-8}$ ⇒ 12% increase from 2 400 to 2 415 K

(d) At 6 000 K, $N^*/N_0 = 0.002\,0$

20-8. Tap water: mean = 0.006_1 standard deviation = 0.005_5
Spiked water: mean = 0.038_4 standard deviation = 0.004_9

(a) (mean spike signal − mean unspiked signal) = m[0.50 ppb]
$(0.038_4 - 0.006_1) = m(0.50 \text{ ppb})$ ⇒ $m = 0.064_6$ ppb^{-1}

(b) Detection limit = $\dfrac{3s}{m}$, where s is the standard deviation of the spiked sample

Detection limit = $\dfrac{3(0.004_9)}{0.064_6 \text{ ppb}^{-1}} = 0.23$ ppb

(c) Limit of quantitation = $\dfrac{10s}{m} = \dfrac{10(0.004_9)}{0.064_6 \text{ ppb}^{-1}} = 0.76$ ppb

20-9. (a) In cells B12:C14 of the spreadsheet, we find

slope = $23.5_9 \pm 0.2_8$ (μg/mL)$^{-1}$

intercept = $7._9 \pm 5._2$

$s_y = 5._3$

(b) Inserting a corrected signal from unknown = 423 − 6 = 417 into cell B20 and $k = 1$ measurements of the unknown in cell B21 gives results in cells B22 and B23:   [K$^+$] = 17.3 ± 0.3 μg/mL.

	A	B	C	D	E	F	G	H
1	Least-Squares Spreadsheet							
2		x =	y =					
3	Highlight cells B12:C14	µg/mL	signal-blank					
4	Type "= LINEST(C4:C7,	5	124					
5	B4:B7,TRUE,TRUE)	10	243					
6	For PC, press	20	486					
7	CTRL+SHIFT+ENTER	30	712					
8	For Mac, press							
9	COMMAND+RETURN							
10								
11		LINEST output:						
12	m	23.5898	7.9153	b				
13	s_m	0.2773	5.2340	s_b				
14	R^2	0.9997	5.3250	s_y				
15								
16	n =	4	B16 = COUNT(B4:B7)					
17	Mean y =	391.250	B17 = AVERAGE(C4:C7)					
18	$\Sigma(x_i - \text{mean } x)^2 =$	368.75	B18 = DEVSQ(B4:B7)					
19								
20	Measured y =	417	Input					
21	k = Number of replicate measurements of y =	1	Input					
22	Derived x =	17.34157	B22 = (B20-C12)/B12					
23	$s_x =$	0.2527	B23=(C14/B12)*SQRT((1/B21)+(1/B16)+((B20-B17)^2)/(B12^2*B18))					

Graph shows $y = 23.59x + 7.92$, with axes $x = \mu g/mL$ and $y = \text{signal - blank}$.

20-10. (a) mol added standard = $(0.001\ 00\ \text{L})(0.030\ 0\ \text{M}) = 3.00 \times 10^{-5}$ mol

$[S]_f = (3.00 \times 10^{-5}\ \text{mol})/(0.010\ 0\ \text{L}) = 3.00 \times 10^{-3}$ M

(b) $\dfrac{[K^+]_i}{[K^+]_f + [S]_f} = \dfrac{[K^+]_i}{\frac{1}{2}[K^+]_i + [3.00 \times 10^{-3}\ \text{M}]} = \dfrac{3.00\ \text{mV}}{4.00\ \text{mV}}$

$\Rightarrow [K^+]_i = 3.60 \times 10^{-3}$ M

20-11. Because the serum was not significantly diluted, $[Na^+]_f = [Na^+]_i$, and the standard addition equation takes the form

$\dfrac{[X]_i}{[X]_f + [S]_f} = \dfrac{[Na^+]_i}{[Na^+]_i + 0.104\ \text{M}} = \dfrac{4.27\ \text{mV}}{7.98\ \text{mV}} \Rightarrow [Na^+]_i = 0.120$ M

20-12. $[Cu^{2+}]_f = [Cu^{2+}]_i \dfrac{V_i}{V_f} = 0.950\ [Cu^{2+}]_i$

$[S]_f = [S]_i \dfrac{V_i}{V_f} = (100.0\ \text{ppm})\left(\dfrac{1.00\ \text{mL}}{100.0\ \text{mL}}\right) = 1.00$ ppm

$\dfrac{[Cu^{2+}]_i}{1.00\ \text{ppm} + 0.950[Cu^{2+}]_i} = \dfrac{0.262}{0.500} \Rightarrow [Cu^{2+}]_i = 1.04$ ppm

20-13. (a) The concentrations of added standard are 0, 10.0, 20.0, 30.0, and 40.0 µg/mL. As an example, in cell B6 of the spreadsheet, 1.00 mL containing $1\,000 \times 10^3$ µg/mL was diluted to 100.0 mL.

$$\text{final concentration} = (1\,000 \times 10^3 \text{ µg/mL})\left(\frac{1.00 \text{ mL}}{100.0 \text{ mL}}\right) = 10.0 \text{ µg/mL}$$

(b) The graph below has an x-intercept of -20.4 µg/mL, which is the concentration of unknown after 10.00 mL has been diluted to 100.0 mL. The original concentration of X is $(20.4 \text{ µg/mL})\left(\frac{100.0 \text{ mL}}{10.00 \text{ mL}}\right) = 204$ µg/mL.

	A	B	C	D
1	Standard addition least squares spreadsheet			
2				
3		x =	y =	
4		µg/mL	signal	
5		0.0	0.163	
6		10.0	0.240	
7		20.0	0.319	
8		30.0	0.402	
9		40.0	0.478	
10	Highlight cells B14:C16			
11	B14:C16 = LINEST(C5:C9,B5:B9,TRUE,TRUE)			
12	CTRL+SHIFT+ENTER (PC) or COMMAND+RETURN (Mac)			
13		LINEST output:		
14		m 0.007920	0.162000	b
15		s_m 0.000060	0.001470	s_b
16		R^2 0.999828	0.001897	s_y
17				
18	x-intercept = -b/m =	-20.45		
19				
20	n =	5	B20 = COUNT(C5:C9)	
21	Mean y =	0.32040	B21 = AVERAGE(C5:C9)	
22	$\Sigma(x_i - \text{mean } x)^2 =$	1000	B22 = DEVSQ(B5:B9)	
23				
24	Std deviation of			
25	x-intercept =	0.324661		
26	B25 = (C16/ABS(B14))*SQRT((1/B20) + B21^2/(B14^2*B22))			

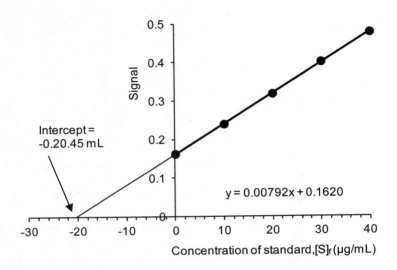

(c) The uncertainty in the intercept is computed in cell B25. The relative uncertainty is 0.325/20.45 = 1.59%. Taking 1.59% of the concentration 204 µg/mL gives an uncertainty of 3 µg/mL. The final answer is 204 ± 3 µg/mL.

20-14. All solutions are made up to the same final volume. Therefore, we plot emission intensity versus concentration of added standard. The graph intercepts the x-axis at −0.164 µg/mL. Because the sample was diluted by a factor of 10, the original sample concentration is 1.64 µg/mL. The relative uncertainty in cells C21 and C20 is 0.004 89/0.163 9 = 2.98%, so the uncertainty in the answer is 2.98% of 1.64 µg/mL = 0.05 µg/mL. Final answer: 1.64 ± 0.05 µg/mL

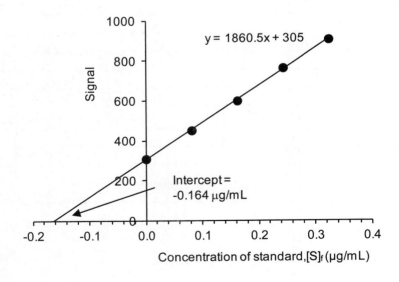

Atomic Spectroscopy

	A	B	C	D
1	Standard addition least squares spreadsheet			
2				
3		x =	y =	
4		µg/mL	signal	
5		0.000	309	
6		0.081	452	
7		0.162	600	
8		0.243	765	
9		0.324	906	
10	Highlight cells B14:C16			
11	B14:C16 = LINEST(C5:C9,B5:B9,TRUE,TRUE)			
12	CTRL+SHIFT+ENTER (PC) or COMMAND+RETURN (Mac)			
13		LINEST output:		
14	m	1860.49	305.00	b
15	s_m	26.31	5.22	s_b
16	R^2	0.9994	6.74	s_y
17				
18	x-intercept = -b/m =	-0.1639		
19				
20	n =		5	B20 = COUNT(C5:C9)
21	Mean y =		606.4	B21 = AVERAGE(C5:C9)
22	$\Sigma(x_i - \text{mean } x)^2 =$		0.0656	B22 = DEVSQ(B5:B9)
23				
24	Std deviation of			
25	x-intercept =	0.0049		
26	B25 = (C16/ABS(B14))*SQRT((1/B20) + B21^2/(B14^2*B22))			

20-15. Standard mixture has 3.42 times as much X as S:

$$\frac{A_X}{[X]} = F\left(\frac{A_S}{[S]}\right) \Rightarrow \frac{0.930}{[3.42]} = F\left(\frac{1.000}{[1.00]}\right) \Rightarrow F = 0.271_9$$

Unknown mixture:

$$\frac{A_X}{[X]} = F\left(\frac{A_S}{[S]}\right) \Rightarrow \frac{1.690}{[X]} = 0.271_9 \left(\frac{1.000}{[0.824 \text{ µg/mL}]}\right) \Rightarrow [X] = 5.12_2 \text{ µg/mL}$$

But X was diluted by a factor of 10.00/50.0, so the original concentration in the unknown was $(5.12_2 \text{ µg/mL})\left(\frac{50.0}{10.00}\right) = 25.6$ µg/mL.

20-16. We find the response factor from the standard mixture of Fe (= X) and Mn (= S):

$$\frac{A_X}{[X]} = F\left(\frac{A_S}{[S]}\right) \Rightarrow \frac{A_{Fe}}{[Fe]} = F\left(\frac{A_{Mn}}{[Mn]}\right)$$

$$\frac{1.05}{[2.50 \text{ µg/mL}]} = F\left(\frac{1.00}{[2.00 \text{ µg/mL}]}\right) \Rightarrow F = 0.840$$

Concentration of Mn in unknown mixture = (13.5 µg/mL) = 2.25 $\frac{\text{µg}}{\text{mL}}$

For the unknown, we can say:

$$\frac{0.185}{[\text{Fe}]} = 0.840 \left(\frac{0.128}{[2.25\ \mu\text{g/mL}]} \right) \Rightarrow [\text{Fe}] = 3.87\ \mu\text{g/mL}$$

Because 5.00 mL of unknown was diluted to 6.00 mL, the original concentration of Fe must have been $\frac{6.00}{5.00}(3.87\ \mu\text{g/mL}) = 4.65\ \mu\text{g/mL} = 8.33 \times 10^{-5}$ M.

20-17. (a) Isobaric interference is caused when an ion in the plasma has nearly the same mass-to-charge ratio as analyte ion. If the mass spectrometer cannot resolve closely spaced masses, the two ions appear to be the same.

(b) $^{40}\text{Ar}^{16}\text{O}^{1}\text{H}^{+}$ has a mass-to-charge ratio of 57, which interferes with $^{57}\text{Fe}^{+}$. $^{32}\text{S}^{16}\text{O}_2^{+}$ interferes with $^{64}\text{Zn}^{+}$, and $^{23}\text{Na}\ ^{35}\text{Cl}^{+}$ interferes with $^{58}\text{Ni}^{+}$.

20-18. (a) CsCl provides Cs atoms, which are easily ionized to $\text{Cs}^{+} + e^{-}$ in the plasma. The electrons in the plasma inhibit the ionization of Sn. Therefore, emission from atomic Sn is not lost to emission from Sn^{+}.

(b)

	A	B	C	D	E	F
1	Tin in canned food - *Anal. Bioanal. Chem.* **2002**, *374*, 235					
2	Calibration data for 189.927 nm					
3						
4	Conc (µg/L)	Signal intensity			Output from LINEST	
5	0	4.0			slope	intercept
6	10	8.5		Parameter	0.781651	0.863321
7	20	19.6		Std Dev	0.018508	1.556732
8	30	23.6		R^2	0.996648	3.213618
9	40	31.1				
10	60	41.7				
11	100	78.8				
12	200	159.1				
13						
14	Select cells E6:F8					
15	Enter the formula = LINEST(B5:B12,A5:A12,TRUE,TRUE)					
16	CONTROL+SHIFT+ENTER on PC or COMMAND+RETURN on Mac					

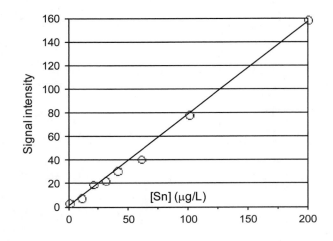

(c) For the 189.927-nm Sn emission line, spike recoveries are all near 100 µg/L, which is near 100%. None of the elements in the table appears to interfere significantly at 189.927 nm. For the 235.485-nm emission line, interference from an emission line of Fe is so serious that the Sn signal cannot be measured. Several other elements interfere enough to reduce the accuracy of the Sn measurement. These elements include Cu, Mn, Zn, Cr, and, perhaps, Mg. The 189.927-nm line is clearly the better of the two wavelengths for minimizing interference.

(d) Limit of detection = minimum detectable concentration = $3s/m$
where s is the standard deviation of the replicate samples and m is the slope of the calibration curve. Putting in the values $s = 2.4$ units and $m = 0.782$ units per (µg/L) gives

$$\text{limit of detection} = \frac{3s}{m} = \frac{3(2.4 \text{ units})}{0.782 \text{ units}/(\mu g/L)} = 9.2 \ \mu g/L$$

$$\text{limit of quantitation} = \frac{10s}{m} = \frac{10(2.4 \text{ units})}{0.782 \text{ units}/(\mu g/L)} = 30.7 \ \mu g/L$$

It would be reasonable to quote a limit of detection as 9 µg/L and a limit of quantitation as 31 µg/L.

(e) A 2-g food sample ends up in a volume of 50 mL. The limit of quantitation is 30.7 µg Sn/L for the solution. A 50-mL volume with Sn at the limit of quantitation contains (0.050 L)(30.7 µg Sn/L) = 1.54 µg Sn. The quantity of Sn per unit mass of food is

$$\frac{(1.54 \ \mu g \ Sn)(1 \ mg/1 \ 000 \ \mu g)}{(2.0 \ g \ food)(1 \ kg/1 \ 000 \ g)} = 0.77 \ \frac{mg \ Sn}{kg \ food}$$

20-19. Standard addition graph: plot signal versus Ti or S concentration.

Ti (ppm)	Signal		S (ppm)	Signal
0.00	0.86		0.0	0.017 4
3.00	1.10		37.0	0.022 1
6.00	1.34		74.0	0.026 8
12.00	1.82		148.0	0.036 2

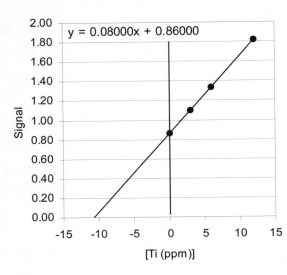 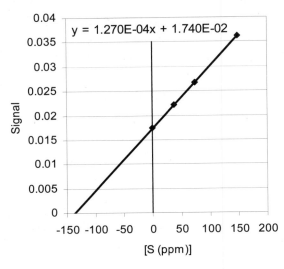

Ti standard addition graph: negative intercept = 0.860/0.080 0 = 10.75 mg/L
S standard addition graph: negative intercept = 0.017 4/0.000 127 = 137.0 mg/L
Ti atomic mass = 47.867   S atomic mass = 32.065
$[Ti] = (10.75 \text{ mg/L})/(47.867 \text{ g/mol}) = 2.246 \times 10^{-4}$ M
$[S] = (137.0 \text{ mg/L})/(32.065 \text{ g/mol}) = 4.273 \times 10^{-3}$ M
$[Transferrin] = [S]/39 = 1.096 \times 10^{-4}$ M
Ti/transferrin = $(2.246 \times 10^{-4} \text{ M})/(1.096 \times 10^{-4} \text{ M}) = 2.05$

20-20. (a) The higher result in sample 2 compared to sample 1 is probably the effect of diluting interfering species so they do not interfere as much in sample 2. Dilution lowers the concentration of species that might react with Li^+ or make smoke that scatters light. In sample 3, interference is present to the same extent as in sample 2, but the standard addition procedure corrects for the interference. The whole point of standard addition is to measure the effect of the complex interfering matrix on the response to known quantities of analyte.

(b) Samples 4–6 use a hotter flame than samples 1–3. High temperature appears to eliminate most of the interference observed at lower temperature. Dilution has only a tiny effect on the results.

(c) Because it appears from samples 1–3 that standard addition gives a true result, we surmise that samples 3 and 6, and possibly 5, are within experimental error of each other. I would probably report the "true" value as the mean of samples 3 and 6 (81.4 ppm). It might also be reasonable to take an average of samples 3, 5, and 6 (80.8 ppm).

CHAPTER 21
PRINCIPLES OF CHROMATOGRAPHY AND MASS SPECTROMETRY

21-1. Eluent is the solvent or gas going into a column. Eluate is what comes out.

21-2. Smaller plate height gives less band spreading: 0.1 mm.

21-3. (a) Material constantly diffuses away from the most concentrated region at the center of the band. This process is called longitudinal diffusion because diffusion occurs in the direction of the length of the column. The longer we wait, the broader the band becomes from diffusion.

(b) If the flow rate is too low, bands spread by longitudinal diffusion. If the flow rate is too high, bands spread because the rate of mass transfer between the mobile and stationary phases is too slow to keep up with the rate at which the mobile phase is moving. In general, there is an intermediate flow rate at which the best trade-off between these two band-broadening mechanisms is obtained.

21-4. Longitudinal diffusion is much faster in a gas than in a liquid, so the flow rate in gas chromatography must be much faster to prevent bands from broadening too much during their passage through the column.

21-5. Silanization caps polar hydroxyl groups with less polar groups. Strong hydrogen bonding between polar solutes and the hydroxyl groups leads to tailing. Eliminating this hydrogen bonding reduces the tailing.

21-6. For qualitative analysis, the mass spectrometer gives the mass of the molecular ion and the molecular masses of characteristic fragment ions. For quantitative analysis, the mass spectrometer detector current is proportional to the quantity of analyte eluted.

21-7. In spiking, an authentic sample of a suspected analyte is added to an unknown to see whether the height of the chromatography peak for the analyte increases. If a new peak appears, the added compound is not the analyte. If the analyte peak grows, the added compound could be the analyte. If the same result is observed in several different columns with different types of stationary phases, it is likely that the added sample is identical to the analyte. We do not expect two different compounds to be retained to the same extent on several different stationary phases because the retention mechanism is different on each stationary phase.

21-8. (a) We need to express t_r and $w_{1/2}$ in the same units, so their units will cancel. Let's convert t_r to seconds:

$$N = \frac{5.55 t_r^2}{w_{1/2}^2} = \frac{5.55(769.8 \text{ s})^2}{(8.7 \text{ s})^2} = 4.3_5 \times 10^4 \text{ plates}$$

(b) $H = (158 \text{ mm})/(4.3_5 \times 10^4 \text{ plates}) = 3.6 \text{ µm/plate}$

21-9. (a) $w_{1/2} = 0.8$ mm for ethyl acetate and 2.6 mm for toluene.

(b) $t_r = 11.3$ mm for ethyl acetate and 36.2 mm for toluene. The number of theoretical plates is $N = 5.55 t_r^2/w_{1/2}^2 = 1.1 \times 10^3$ for ethyl acetate and 1.1×10^3 for toluene.

21-10. $\dfrac{\text{large load}}{\text{small load}} = \left(\dfrac{\text{large column radius}}{\text{small column radius}}\right)^2$

$\dfrac{250 \text{ mg}}{12 \text{ mg}} = \left(\dfrac{\text{large column radius}}{0.75 \text{ cm}}\right)^2$

$\Rightarrow$ large column radius = 3.4_2 cm

The large column should be approximately 6.8 cm diameter × 25 cm long. The volume flow rate should be increased in proportion to the cross-sectional area ($\propto r^2$). Therefore, the flow rate for the 250-mg sample should be $\left(\dfrac{3.4_2 \text{ cm}}{0.75 \text{ cm}}\right)^2$ times greater than the flow rate for the 12-mg sample = $\left(\dfrac{3.4_2 \text{ cm}}{0.75 \text{ cm}}\right)^2 (0.8 \text{ mL/min}) = $ 17 mL/min.

21-11. (a) $w_{1/2} = 0.172$ min $\Rightarrow N = \dfrac{5.55 \, t_r^2}{w_{1/2}^2} = \dfrac{5.55(13.81 \text{ min})^2}{(0.172 \text{ min})^2} = 3.58 \times 10^4$ plates

(b) H = plate height = $(30.0 \times 10^3 \text{ mm})/(3.58 \times 10^4 \text{ plates}) = 0.838$ mm/plate

(c) w (measured) = 0.311 min
$w/w_{1/2}$ (measured) = $(0.311 \text{ min})/(0.172) = 1.81$

21-12. (a) $w_{1/2}$ (heptane) = 0.126 min; $w_{1/2}$ ($C_6H_4F_2$) = 0.119 min

$N = \dfrac{5.55 \, t_r^2}{w_{1/2}^2} = \dfrac{5.55(14.56 \text{ min})^2}{(0.126 \text{ min})^2} = 7.41 \times 10^4$ plates for heptane

H = plate height = 30.0 m/7.41×10^4 plates = 0.404 mm/plate

$N = \dfrac{5.55(14.77 \text{ min})^2}{(0.119 \text{ min})^2} = 8.55 \times 10^4$ plates for $C_6H_4F_2$

H = plate height = $30.0 \text{ m}/8.55 \times 10^4 \text{ plates} = 0.351 \text{ mm/plate}$

(b) w (heptane) = 0.214 min; w ($C_6H_4F_2$) = 0.202 min

$$w_{av} = \frac{1}{2}(0.214 + 0.202) = 0.208 \text{ min}$$

$$\text{resolution} = \frac{\Delta t_r}{w_{av}} = \frac{14.77 - 14.56}{0.208} = 1.01$$

21-13. (a) $\dfrac{\text{Large load}}{\text{Small load}} = \left(\dfrac{\text{large column radius}}{\text{small column radius}}\right)^2$

$\dfrac{72.4 \text{ mg}}{10.0 \text{ mg}} = \left(\dfrac{1.50 \text{ cm}}{\text{small column radius}}\right)^2 \Rightarrow \text{radius} = 0.557 \text{ cm}$

$\Rightarrow \text{diameter} = 1.11 \text{ cm}$ (Length is unchanged at 32.6 cm.)

(b) Scale volume in proportion to mass of analyte.

$\Rightarrow \text{volume} = \left(\dfrac{10.0 \text{ mg}}{72.4 \text{ mg}}\right)(0.500 \text{ mL}) = 0.069 \text{ mL}$

(c) $\dfrac{\text{Large flow rate}}{\text{Small flow rate}} = \left(\dfrac{\text{large column radius}}{\text{small column radius}}\right)^2 = \left(\dfrac{1.50 \text{ cm}}{0.557 \text{ cm}}\right)^2$

Putting in large flow rate = 1.85 mL/min gives small flow rate = 0.256 mL/min.

21-14. 50 cm of column with a diameter of 0.25 mm has a volume of $\pi r^2 \times \text{length} = \pi(0.0125 \text{ cm})^2 \times 50 \text{ cm} = 0.0245 \text{ cm}^3 = 0.0245 \text{ mL}$. A flow rate of 0.0245 mL/s corresponds to (0.0245 mL/s)(60 s/min) = 1.47 mL/min.

21-15. (a) $\dfrac{A_X}{[X]} = F\left(\dfrac{A_S}{[S]}\right) \Rightarrow \dfrac{3\,473}{3.47 \text{ mM}} = F\left(\dfrac{10\,222}{1.72 \text{ mM}}\right) \Rightarrow F = 0.168_4$

(b) $[S] = \dfrac{(8.47 \text{ mM})(1.00 \text{ mL})}{10.0 \text{ mL}} = 0.847 \text{ mM}$

(c) $\dfrac{A_X}{[X]} = F\left(\dfrac{A_S}{[S]}\right) \Rightarrow \dfrac{5\,428}{[X]} = 0.168_4\left(\dfrac{4\,431}{0.847 \text{ mM}}\right) \Rightarrow [X] = 6.16 \text{ mM}$

(d) The concentration of X in the original unknown was twice that of the diluted concentration, so [X] in original unknown = 12.3 mM.

21-16. First find the response factor from the known mixture:

$\dfrac{A_C}{[C]} = F\left(\dfrac{A_D}{[D]}\right) \Rightarrow \dfrac{4.42}{236 \text{ µg/mL}} = F\left(\dfrac{5.52}{337 \text{ µg/mL}}\right) \Rightarrow F = 1.14_{34}$

Then find the concentration of C in the unknown mixture:

$$\frac{3.33}{[C]} = 1.143_4 \left(\frac{2.22}{[1\,230\ \mu g/25.00\ mL]} \right) \Rightarrow [C] = 64.5_4\ \mu g/mL$$

$$[C]\text{ in original unknown} = \frac{(25.00\ mL)(64.5_4\ \mu g/mL)}{10.00\ mL} = 161\ \mu g/mL$$

21-17.

Sample	x = concentration ratio $C_{10}H_8/C_{10}D_8$	y = area ratio $C_{10}H_8/C_{10}D_8$
1	0.10	0.101
2	0.50	0.573
3	1.00	1.072

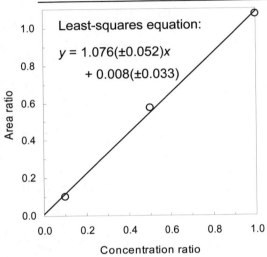

Least-squares equation:
$y = 1.076(\pm 0.052)x + 0.008(\pm 0.033)$

The three points lie on a straight line with a slope of $F = 1.08 \pm 0.05$ and an intercept of 0.008 ± 0.033. The intercept is not significantly different from zero because the standard deviation of the intercept is larger than the intercept.

	A	B	C	D
1	Least-Squares Spreadsheet			
2				
3	Highlight cells B10:C12	x	y	
4	Type "= LINEST(C4:C6,	0.1	0.101	
5	B4:B6,TRUE,TRUE)	0.5	0.573	
6	For PC, press	1	1.072	
7	CTRL+SHIFT+ENTER			
8	For Mac, press			
9	APPLE+ENTER	LINEST output:		
10		m 1.07557	0.00836	b
11		s_m 0.05168	0.03349	s_b
12		R^2 0.99770	0.03296	s_y

21-18. (a) The spreadsheet and graph below show that minimum plate height occurs at a flow rate of 31.6 mL/min.

	A	B	C
1	van Deemter calculations [H = A + B/u + Cu]		
2			
3	A(mm) =	Flow rate (u, mL/min)	H (mm)
4	1.5	2	14.050
5	B(mm*mL/min) =	5	6.625
6	25	10	4.250
7	C(mm*min/mL) =	20	3.250
8	0.025	30	3.083
9		31.6	3.081
10		40	3.125
11		80	3.813
13		120	4.708
14	C4 = $A$4 + $A$6/B4 + $A$8*B4		

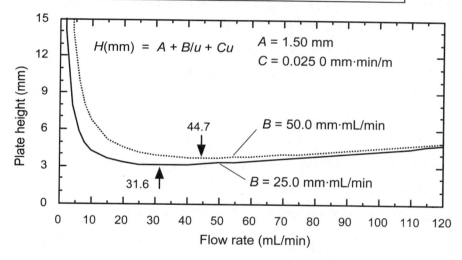

(From calculus, the minimum occurs when $dH/du = 0$.

$dH/du = -B/u^2 + C = 0 \Rightarrow$

$$u_{optimum} = \sqrt{\frac{B}{C}} = \sqrt{\frac{25.0}{0.025\,0}} = 31.6 \text{ mL/min}$$

(b) If the rate of longitudinal diffusion is increased, we must run the column faster so that the bands do not spread too much. The optimum flow rate should increase and the optimum plate height will be larger (worse) than with the slower longitudinal diffusion. The graph above shows that doubling B increases the optimum flow rate to 44.7 mL/min.

(c) If the rate of equilibration between phases is increased, we can run the column faster without spreading the bands too much. The optimum flow rate should increase and the optimum plate height will be smaller (better) than with the slower equilibration between phases.

21-19. (a) $H(\text{mm}) = A + B/u + Cu$

$= (1.50 \text{ mm}) + \dfrac{25.0 \text{ mm·mL/min}}{20.0 \text{ mL/min}} + (0.025\ 0 \text{ mm·min/mL})(20.0 \text{ mL/min})$

$= 3.25 \text{ mm}$

(b) $H = L/N$ (L = column length, N = number of plates)

$N = L/H = 2.00 \text{ m}/0.003\ 25 \text{ m} = 615$ plates

(c) $N = \dfrac{5.55 t_r^2}{w_{1/2}^2} \Rightarrow w_{1/2} = t_r \sqrt{\dfrac{5.55}{N}} = (8.00 \text{ min})\sqrt{\dfrac{5.55}{615}} = 0.760$ min

21-20. *Molecular mass* is a weighted average mass based on the natural abundance of all isotopes in the molecular formula. To compute molecular mass, we sum the atomic masses listed in the periodic table, which are weighted averages of the isotopes of each element. *Nominal mass* is the integer mass of the species with the most abundant of each of the isotopes in the formula.

C_6H_6 molecular mass $= 6 \times 12.010\ 7 + 6 \times 1.007\ 94 = 78.111\ 8$

C_6H_6 nominal mass $= 6 \times 12 + 6 \times 1 = 78$

21-21. Atomic mass of Cl = Σ(atomic mass × natural abundance)

$= (34.968\ 85)(0.757\ 8) + (36.965\ 90)(0.242\ 2) = 35.453$

Happily, the value in the periodic table is also 35.453.

21-22. $C_4H_{11}N_3S^+$:

4 $^{12}C = 4 \times 12 =$	+48 (exact)
11 $^1H = 11 \times 1.007\ 825 =$	+11.086 075
3 $^{14}N = 3 \times 14.003\ 07 =$	+42.009 21
$^{32}S =$	+31.972 07
subtract 1 electron	− 0.000 549
sum	133.066 81

$C_4H_{11}N_3O_2^+$:

4 $^{12}C = 4 \times 12 =$	+48 (exact)
11 $^1H = 11 \times 1.007\ 825 =$	+11.086 075
3 $^{14}N = 3 \times 14.003\ 07 =$	+42.009 21
2 $^{16}O = 2 \times 15.994\ 91$	+31.989 82
subtract 1 electron	− 0.000 549
sum	133.084 56

The observed mass of 133.068 6 is consistent with $C_4H_{11}N_3S^+$.

21-23. (a)

Propazine = $C_9H_{16}N_5Cl$

PropazineH$^+$ is $C_9H_{17}N_5Cl^+$ with H$^+$ on one of the two N atoms outside the ring. ^{35}Cl is the predominant isotope of Cl.

9 ^{12}C = 9 × 12 =	+108 (exact)
17 ^1H = 17 × 1.007 825 =	+17.133 025
5 ^{14}N = 5 × 14.003 07 =	+70.015 35
1 ^{35}Cl = 1 × 34.968 85	+34.968 85
subtract 1 electron	− 0.000 549
sum	230.116 68 (observed m/z = 230.116 4)

(b) The ion at m/z 231 predominantly has one ^{13}C in place of one ^{12}C because the natural abundance of ^{13}C (1.07%) is substantially higher than the abundance of ^2H (0.012%) or ^{15}N (0.368%).

m/z 231 is mainly $^{12}C_8\,^{13}CH_{17}N_5Cl^+$

8 ^{12}C = 8 × 12	=	+96 (exact)
1 ^{13}C = 1 × 13.003 35	=	+13.003 35
17 ^1H = 17 × 1.007 825	=	+17.133 025
5 ^{14}N = 5 × 14.003 07	=	+70.015 35
1 ^{35}Cl = 1 × 34.968 85	=	+34.968 85
subtract 1 electron		− 0.000 549
sum		231.120 03 (observed m/z = 231.118 8)

The observed mass is low because some of the observed peak comes from $^{12}C_9H_{17}\,^{14}N_4\,^{15}N\,Cl^+$, which has a lower mass.

Expected intensity = 9 × 1.1% = 9.9% (observed intensity = 11.7%)

(c) m/z 232 is mainly $^{12}C_9H_{17}N_5\,^{37}Cl^+$

9 ^{12}C = 9 × 12	=	+108 (exact)
17 ^1H = 17 × 1.007 825	=	+17.133 025
5 ^{14}N = 5 × 14.003 07	=	+70.015 35
1 ^{37}Cl = 1 × 36.965 90	=	+36.965 90
subtract 1 electron		− 0.000 549
sum		232.113 73 (observed m/z = 232.113 4)

21-24. $31 = CH_2OH^+$; $41 = C_3H_5^+$; $43 = C_3H_7^+$; $56 = C_4H_8^+$ (loss of H_2O)

21-25. (a) C atoms $\approx 5.9/1.1 = 5.4 \approx 5$. There must be an odd number of N atoms. A possible formula is C_5H_5N.

(b) C atoms $\approx 6.1/1.1 = 5.5 \approx 5$ or 6. There must be an odd number of N atoms. Possible formulas are $C_5H_5ON_3$ and $C_6H_5O_2N$.

(c) C atoms $\approx 7.4/1.1 = 6.7 \approx 7$. There must be an even number of N atoms. Possible formulas with 7 carbons include $C_7H_4O_2N_2$ and C_7O_4 (an exotic molecule). The $(M+1)/M^+$ ratio 7.4% could possibly represent 6 carbon atoms. Possible formulas with 6 carbons include $C_6H_{12}O_4$ and $C_6H_{16}O_2N_2$.

(d) C atoms $\approx 12.5/1.1 = 11.4 \approx 11$ or 12. There must be an even number of N atoms. Possible formulas with 11 C are $C_{11}H_{20}O$, $C_{11}H_4O_2$, and $C_{11}H_8N_2$. Possible formulas with 12 C are $C_{12}H_{24}$ and $C_{12}H_8O$.

21-26. The only hydrogen isotope to consider is ^1H, because ^2H has a natural abundance of 0.012% and we are disregarding contributions <0.1%. Chlorine has 2 isotopes with abundances of $^{35}Cl = 75.78\%$ and $^{37}Cl = 24.22\%$. The ion at m/z 36 is $^1H^{35}Cl^+$ with an abundance of 0.757 8. Disregarding ^2H, we do not expect any intensity at m/z 37. The ion at m/z 38 is $^1H^{37}Cl^+$ with an abundance of 0.242 2. The intensity ratio 38/36 should be 0.242 2/0.757 8 = 0.319 6. If we set the intensity at m/z 36 equal to 100, then the relative intensities are

Intensity ratio 36 : 37 : 38 = 100 : 0 : 31.96

21-27. The only hydrogen isotope to consider is ^1H, because ^2H has a natural abundance of 0.012% and we are disregarding contributions <0.1%. Sulfur has 3 significant isotopes with abundances of $^{32}S = 94.93\%$, $^{33}S = 0.76\%$, and $^{34}S = 4.29\%$. The molecular ion at m/z 34 is $^1H_2{}^{32}S^+$, and it has an abundance of 0.949 3. The ion at m/z 35 is $^1H_2{}^{33}S^+$, with an abundance of 0.007 6. The ion at m/z 36 is $^1H_2{}^{34}S^+$, with an abundance of 0.042 9. We normalize to an intensity of 100 at m/z 34 by dividing each abundance by 0.949 3 and multiplying by 100:

Intensity ratio 34 : 35 : 36 = 100 : 0.80 : 4.52

21-28. The observed mass spectrum in the region of the molecular ion is shown below.

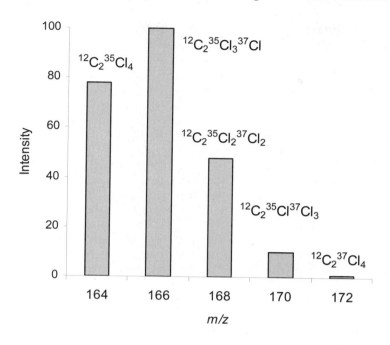

The peak at m/z 164 corresponds to the isotopic composition $^{12}C_2{}^{35}Cl_4$. The relative abundances of ^{35}Cl and ^{37}Cl are 0.757 8 and 0.242 2. The most probable combination of isotopes, which gives the tallest peak, is $^{12}C_2{}^{35}Cl_3{}^{37}Cl$, with m/z = 166. Isotopic molecules $^{12}C_2{}^{35}Cl_2{}^{37}Cl_2$, $^{12}C_2{}^{35}Cl{}^{37}Cl_3$, and $^{12}C_2{}^{37}Cl_4$ have successively lower abundance. If all peaks had been listed in the problem, we would also find weak signals for species with one ^{12}C and one ^{13}C, such as $^{12}C^{13}C\,{}^{35}Cl_3{}^{37}Cl$, with m/z = 167.

21-29. The general procedure is to inject a fixed amount of $^{13}C_3$-caffeine internal standard into coffee and extract both the ordinary caffeine and the internal standard by solid-phase microextraction. Following extraction, insert the fiber into a gas chromatograph equipped with a mass spectrometer for a detector. Caffeine is detected by setting the mass spectrometer to measure mass 194 only. The internal standard is measured by monitoring mass 197 only. Even though the two substances are eluted at exactly the same position, the mass spectrometer can be set to measure just the analyte or just the internal standard.

The response factor (which ought to be close to 1.00) can be established by making a known solution of $^{13}C_3$-caffeine and $^{12}C_3$-caffeine, extracting with the fiber, and recording the gas chromatogram. An even better idea is to establish a calibration curve from a series of aqueous solutions containing a fixed amount of

$^{13}C_3$-caffeine and variable amounts of ordinary $^{12}C_3$-caffeine comparable to the quantities expected in coffee. A graph of peak area ratio ($^{12}C_3$-caffeine/$^{13}C_3$-caffeine) versus concentration ratio ($^{12}C_3$-caffeine/$^{13}C_3$-caffeine) should be a straight line.

CHAPTER 22
GAS AND LIQUID CHROMATOGRAPHY

22-1. (a) Wall-coated: liquid stationary phase bonded to the wall of column
Support-coated: liquid stationary phase on solid support on wall of column
Porous-layer: solid stationary phase on wall of column

(b) The bonded stationary phase does not bleed from the column during use.

(c) Makeup gas is added after the column and before the detector to provide the optimum type of gas and optimum flow rate for a particular detector.

(d) In solvent trapping, the initial column temperature is low enough to condense solvent at the beginning of the column. Solute is very soluble in the solvent and is trapped in a narrow band at the start of the column. In cold trapping, the initial column temperature is 150° lower than the boiling points of solutes, which condense in a narrow band at the start of the column. In both cases, elution occurs as the column temperature is raised.

22-2. (a) At very low flow rate, solute bands have time to spread by longitudinal diffusion because they spend such a long time in the column. Band broadening is synonymous with increased plate height.

At very high flow rate, solute cannot equilibrate fast enough between the mobile and stationary phases. Some solute in the stationary phase is left behind solute in the mobile phase, so the band spreads.

(b) Given that we do not observe significant band spreading at high flow rate for the smallest particle size, we conclude that solute equilibrates rapidly between the mobile and stationary phases. That is, solute has adequate time to go between phases over the range of flow rates in the graph. Presumably, at a sufficiently high flow rate, we would observe band broadening.

22-3. The thermal conductivity detector measures changes in the thermal conductivity of the gas stream exiting the column. Any substance other than the carrier gas will change the conductivity of the carrier gas. Therefore, the detector responds to all analytes. The flame ionization detector burns eluate in a H_2/O_2 flame to create CH radicals from carbon atoms (except carbonyl and carboxyl carbons), which then go on to be ionized to a small extent in the flame: $CH + O \rightarrow CHO^+ + e^-$. Most other kinds of molecules do not create ions in the flame and are not detected.

22-4. (a) A thin stationary phase permits rapid equilibration of analyte between the mobile and stationary phases, reducing C in the van Deemter equation. A thin stationary phase in a narrow-bore column gives small plate height and high resolution. In a wide-bore column, the thick film requires more time for solutes to diffuse through the film and therefore more time for equilibration between the stationary and mobile phases. Also, molecules in the mobile phase require more time in a wide-diameter column than in a small-diameter column to diffuse across the diameter of the column.

(b) *Narrow-bore column:* The column has 5 000 plates per meter, so plate height = $1/(5\,000\ \text{m}^{-1}) = 2.0 \times 10^{-4}$ m = 200 μm. The area of a length (ℓ) of the inside wall of the column is $\pi d \ell$, where d is the column diameter. The volume of stationary phase in this length is $\pi d \ell t$, where t is the thickness of the stationary phase. For $d = 0.25$ mm $= 2.5 \times 10^{-4}$ m, $\ell = 200$ μm $= 2.0 \times 10^{-4}$ m, and $t = 0.10$ μm $= 1.0 \times 10^{-7}$ m, the volume is $1.5_7 \times 10^{-14}$ m^3 = $1.5_7 \times 10^{-8}$ cm^3 = $1.5_7 \times 10^{-8}$ mL. With a density of 1.0 g/mL, the mass of stationary phase in one theoretical plate is $1.5_7 \times 10^{-8}$ g.
Wide-bore column: For $d = 5.3 \times 10^{-4}$ m, $\ell = 6.6_7 \times 10^{-4}$ m, and $t = 5.0 \times 10^{-6}$ m, the volume is $5.5_5 \times 10^{-12}$ m^3 = $5.5_5 \times 10^{-6}$ mL. Mass of stationary phase is $(5.5_5 \times 10^{-6}\ \text{mL})(1.0\ \text{g/mL}) = 5.5_5 \times 10^{-6}$ g.

(c) For the narrow-bore column, 1.0 % of the mass of stationary phase is $(0.010)(1.5_7 \times 10^{-8}\ \text{g}) = 1.5_7 \times 10^{-10}$ g = 0.16 ng.
For the wide-bore column 1.0 % of the mass of stationary phase is $5.5_5 \times 10^{-8}$ g = 56 ng.

22-5. (a) Number of plates: $N = \dfrac{5.55 t_r^2}{w_{1/2}^2} = \dfrac{5.55(49.4\ \text{min})^2}{0.26\ \text{min}^2} = 2.0 \times 10^5$

Plate height: $H = L/N = (100\ \text{m})/(2.0 \times 10^5) = 0.50$ mm

Resolution = $\dfrac{0.589 D t_r}{w_{1/2\text{av}}} = \dfrac{0.589\ (0.55\ \text{min})}{0.25\ \text{min}} = 1.3$

(b) Gas chromatography plate height/liquid chromatography plate height ≈ 500 μm/10 μm = 50 times greater. Plate height is so much greater for gas chromatography because longitudinal diffusion of molecules away from the center of the band is much faster in the gas phase than in the liquid phase.

22-6. The spreadsheet gives $n = 12.44$ in cell F12. There are probably 12 or 13 CH_2 groups in the unknown compound.

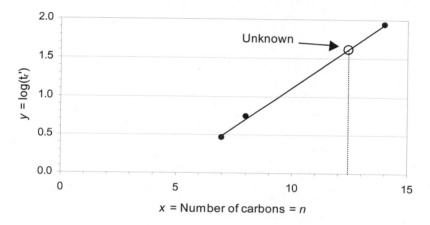

22-7. (a) Selected ion monitoring is measuring ion abundance for m/z 73. Only three compounds in the extract have appreciable intensity at m/z 73.

(b) The base peak for both MTBE and TAME is at m/z 73. This mass corresponds to M-15 (loss of CH_3) for MTBE and M-29 (loss of C_2H_5) for TAME. Loss of the ethyl group bound to carbon in TAME suggests that the methyl group lost from MTBE is also bound to carbon, not to oxygen. If the methyl group bound to oxygen were easily lost from MTBE and TAME, we would expect to see the ethyl group bound to oxygen lost from ETBE. There is no significant peak at M-29 (m/z 73) in ETBE. The following structures are suggested:

MTBE	ETBE	TAME

$$\text{MTBE} \quad \overset{+\cdot}{-\text{O}}\!\!+\!\!\!+ \;\longrightarrow\; -\text{O}\!\!+\!\!\!+ \; + \quad m/z\ 73$$

$$\text{ETBE} \quad \sim\!\!\overset{+\cdot}{\text{O}}\!\!+\!\!\!+ \;\longrightarrow\; \sim\!\!\text{O}\!\!+\!\!\!+ \; + \quad m/z\ 87$$

$$\text{TAME} \quad -\overset{+\cdot}{\text{O}}\!\!+\!\!\!\!\sim \;\longrightarrow\; -\text{O}\!\!+\!\!\!+ \; + \quad m/z\ 73 \qquad\qquad -\text{O}\!\!\overset{+}{+}\!\!\sim \quad m/z\ 87$$

22-8. (a) Caffeine has a nominal mass of 194. The transition $m/z\ 194 \to 109$ represents loss of $C_3H_3NO_2$ (85 mass units), probably as follows:

[structure of caffeine cation with indicated cleavage]

$(^{13}CH_3)_3$-caffeine has a nominal mass of 197. When the $C_3H_3NO_2$ fragment is lost, it carries away one ^{13}C. The transition for $(^{13}CH_3)_3$-caffeine is therefore a loss of 86 mass units, or $m/z\ 197 \to 111$.

(b) An isotopic variant of the analyte is a nearly ideal internal standard because the isotopic variant has nearly the same physical properties as the analyte. It should behave in nearly the same manner as analyte during all steps in sample preparation and analysis.

(c) Caffeine is $C_8H_{10}N_4O_2$ (194 Da). The peak at $m/z\ 195$ is $^{13}C^{12}C_7H_{10}N_4O_2$ containing one ^{13}C anywhere in the molecule. Because the natural abundance of ^{13}C is 1.1%, we expect that a compound with 8 carbon atoms will have an isotopic peak at M+1 with a relative abundance of $(M+1)/M^+ = 8 \times 1.1\% = 8.8\%$. The observed intensity is 10.3%.

22-9. (a) Because nonpolar compounds are more soluble in methanol than in water, the retention time will be shorter in 90% methanol.

(b) The more polar the solvent, the greater the eluent strength on a polar column and the shorter the retention time. Because acetonitrile is less polar than water, 60 vol% acetonitrile is less polar than 40 vol% acetonitrile. The retention time will be greater for the 60 vol% eluent.

(c) In hydrophobic interaction chromatography, retention time increases as the fraction of organic solvent is increased.

(d) In normal-phase chromatography, eluent strength increases as solvent is made more polar. 2-Propanol is more polar than methyl *t*-butyl ether, so retention times will decrease as the fraction of 2-propanol is increased.

22-10. Nonpolar bonded phases:

Silica particle—Si—O—Si(CH$_3$)$_2$—(CH$_2$)$_{17}$CH$_3$

Silica particle—Si—O—Si(CH$_3$)$_2$—(CH$_2$)$_3$—C$_6$H$_5$

Polar bonded phases:

Silica particle—Si—O—Si(CH$_3$)$_2$—(CH$_2$)$_3$—NH$_2$

Silica particle—Si—O—Si(CH$_3$)$_2$—CH$_2$—O—CH$_2$—CH(OH)—CH$_2$OH

22-11. (a) Small particles give increased resistance to flow. High pressure is required to obtain a usable flow rate.

(b) Efficiency increases because solute equilibrates between phases more rapidly if the distance through which the solute must diffuse decreases. Also, migration paths between small particles are more uniform, decreasing broadening from multiple flow paths.

(c) Sub-2-μm particles (UPLC) provide higher resolution. Alternatively, we can obtain the same kind of resolution available with larger particles, but run the separation faster. The principal disadvantage is that UPLC requires much higher pressure than do larger particles.

(d) Superficially porous particles provide the same kind of resolution as sub-2-μm particles, but without requiring such high pressure for operation.

22-12. Plates = (15 cm)/(10.0 × 10⁻⁴ cm/plate) = 1.5×10^4

$$N = \frac{5.55 t_r^2}{w_{1/2}^2} \Rightarrow w_{1/2} = t_r\sqrt{\frac{5.55}{N}} = (10.0 \text{ min})\sqrt{\frac{5.55}{1.5 \times 10^4}} = 0.19_2 \text{ min}$$

If plate height = 5.0 µm, plates = 3.0×10^4 and $w_{1/2} = 0.13_6$ min

22-13. (a) Number of plates: $N = \dfrac{5.55 t_r^2}{w_{1/2}^2} = \dfrac{5.55[(0.63 \text{ min})(60 \text{ s/min})]^2}{(2.3 \text{ s})^2} = 1\,500$

Plate height: $H = L/N = (50 \text{ mm})/(1\,500) = 0.033 \text{ mm} = 33$ µm/plate

Particles in one plate = (33 µm per plate)/(1.7 µm per particle) ≈ 20

(b) An optimum plate height of 4 µm is spanned by (4 µm per plate)/(1.7 µm per particle) ≈ 2 particles. The column in part (a) is being run for maximum speed at a substantial sacrifice in resolution.

22-14. (a) At pH 3, the predominant forms are neutral RCO_2H and cationic RNH_3^+.

(b) The amine will be eluted first, because the cation RNH_3^+ is less soluble than the neutral RCO_2H in the nonpolar stationary phase.

22-15. (a) We need to lower the eluent strength. Use a lower percentage of acetonitrile to increase the retention times and probably increase the resolution.

(b) We need to lower the eluent strength. In normal-phase chromatography, solvent strength increases as the solvent becomes more polar, which corresponds to increasing the methyl *t*-butyl ether concentration. We need a higher concentration of hexane to lower the solvent strength, increase the retention times, and probably increase the resolution.

22-16. Use a slower flow rate, a longer column, or smaller particle size.

22-17. (a)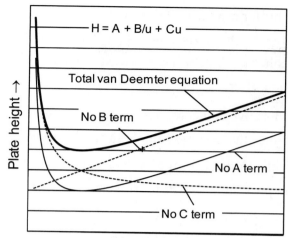

B term: longitudinal diffusion

C term: finite rate of equilibrium (also called mass transfer)

(b) If a column is run at very high speed, then the C term (finite rate of equilibrium) becomes a significant contributor to band broadening.

22-18. First find the response factor from the known mixture:

$$\frac{A_C}{[C]} = F\left(\frac{A_D}{[D]}\right) \Rightarrow \frac{10.86}{1.03 \text{ mg/mL}} = F\left(\frac{4.37}{1.16 \text{ mg/mL}}\right) \Rightarrow F = 2.79_9$$

Then find the concentration of C in the unknown mixture:

$$\frac{5.97}{[C]} = 2.79_9 \left(\frac{6.38}{[12.49 \text{ mg}/25.00 \text{ mL}]}\right) \Rightarrow [C] = 0.167 \text{ mg/mL}$$

$$[C] \text{ in original unknown} = \frac{25.0}{10.0}(0.167 \text{ mg/mL}) = 0.418 \text{ mg/mL}$$

22-19. (a) Spherical volume = $\frac{4}{3}\pi r^3 = \frac{4}{3}\pi(5.00 \times 10^{-4} \text{ cm})^3 = 5.24 \times 10^{-10} \text{ cm}^3$

Mass of one sphere = $(5.24 \times 10^{-10} \text{ mL})(2.20 \text{ g/mL}) = 1.15 \times 10^{-9}$ g

Number of particles in 1.00 g = $1.00 \text{ g}/(1.15 \times 10^{-9} \text{ g/particle}) = 8.68 \times 10^8$

(b) Surface area of one particle = $4\pi r^2 = 4\pi(5.00 \times 10^{-6} \text{ m})^2 = 3.14 \times 10^{-10} \text{ m}^2$

Surface area of 8.68×10^8 particles = 0.273 m^2

(c) Because the observed surface area is 300 m², the particles must be very porous, with a great deal of internal surface area. The internal surface area is about 1 000 times greater than the calculated external surface.

22-20.
1. We need to remove buffer salts from the column before washing with strong (mostly organic) solvent. Otherwise, the salts could precipitate out of solution inside the column. A good way to wash salts away is to use aqueous solvent containing no buffer.

2. Strong solvent will dissolve and carry away from the stationary phase strongly retained solutes that are not soluble in weak solvent.

3. Storing overnight with strong solvent dissolves nonpolar solutes from the stationary phase and prevents them from irreversibly adhering to stationary phase.

4. When you are ready to resume chromatography, the stationary phase must equilibrate with the new mobile phase.

5. Before chromatography of an unknown, it is a good idea to run a standard mixture and make sure that the retention times and plate count behave as expected. If they do not, the column is not ready to use.

22-21. (a)

CocaineH$^+$
$C_{17}H_{22}NO_4$
m/z 304

$C_{10}H_{16}NO_2$
m/z 182

$C_6H_5CO_2$ has a mass of 121 Da. Subtracting 121 Da from 304 Da gives 183 Da. The peak at m/z 182 probably represents cocaineH$^+$ minus $C_6H_5CO_2H$. The structure might be as shown above or some rearranged form of it.

(b) The ion at m/z 304 was selected by mass filter Q1. Its isotopic partner containing ^{13}C at m/z 305 was blocked by Q1. Because the species at m/z 304 is isotopically pure, there is no ^{13}C-containing partner for the collisionally activated dissociation product at m/z 182.

(c) In selected reaction monitoring, mass filter Q1 selects just m/z 304, which eliminates components of plasma that do not give a signal at m/z 304. Then this ion is passed to the collision cell in which it breaks into a major fragment at m/z 182, which passes through Q3. Few, if any, other components in plasma that give a signal at *m/z* 304 also break into a fragment at *m/z* 182. The 2-step selection process eliminates everything else in the sample and produces just one clean peak in the chromatogram.

(d) The phenyl group must be labeled with deuterium because the labeled product gives the same fragment at *m/z* 182 as unlabeled cocaine.

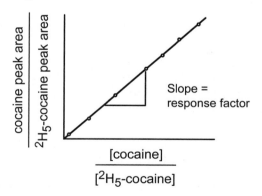

(e) First construct a calibration curve to get the response factor for cocaine compared to $^2H_5$-cocaine. Prepare a series of solutions with known concentration ratios [cocaine]/[$^2H_5$-cocaine] and measure the area of each chromatographic peak in the chromatography/atmospheric chemical ionization/selected reaction monitoring experiment. A graph would be constructed in which (peak area of cocaine)/(peak area of $^2H_5$-cocaine) is plotted versus [cocaine]/[$^2H_5$-cocaine]. The slope of this line is the response factor.

For quantitative analysis, inject a known amount of internal standard $^2H_5$-cocaine into the blood plasma. From the calibration curve, the relative peak areas tell us the relative concentrations of cocaine and the internal standard. Knowing the quantity of internal standard injected into the plasma, we can therefore calculate the quantity of cocaine.

22-22.

HO—C₆H₄—C(CH₃)₂—C₆H₄—O⁻   →   ⁻O—C₆H₄—C(=CH₂)CH₃

m/z 227 C₁₅H₁₅O₂   m/z 133 C₉H₉O

22-23. The idea of purge and trap is to collect all of the analyte from the unknown and to inject all of the analyte into the chromatography column. Splitless injection is required so that analyte is not lost during injection. Any unknown loss of analyte would lead to an error in quantitative analysis.

22-24. Derivatization uses a chemical reaction to convert analyte into a form that is more convenient to separate or easier to detect. For example, RCH_2OH is converted to $RCH_2OSi(CH_3)_3$ to make it more volatile for gas chromatography and to give characteristic mass spectral peaks that aid in identification.

22-25. In solid-phase microextraction, analyte is extracted from a liquid or gas into a thin coating on a silica fiber extended from a syringe. After extraction, the fiber is withdrawn into the syringe. To inject the sample into a chromatograph, the metal needle is inserted through the septum and then the fiber is extended into the injection port. Analyte slowly evaporates from the fiber in the high-temperature port. Cold trapping is required to condense analyte at the start of the column during the slow evaporation from the fiber. If cold trapping were not used, the peaks would be extremely broad because of the slow evaporation from the fiber. During solid-phase microextraction, analyte equilibrates between the unknown and the coating on the fiber. Only a fraction of analyte is extracted into the fiber.

22-26. A molecularly imprinted polymer is created from monomers in the presence of analyte as a template. The template organizes monomers to interact with analyte prior to polymerization. When template is removed from the polymer, the void that is left is the right size for analyte and has functional groups oriented in the correct manner to bind to analyte.

22-27. (a) The vial was heated to increase the vapor pressure of the analyte and the internal standard so there would be enough in the gas phase (the headspace) to extract a significant quantity with the microextraction fiber.

(b) At 60°C, the analyte and internal standard are cold trapped at the beginning of the column. Because desorption from the fiber takes many minutes, we do not want chromatography to begin until desorption is complete.

(c) H—N⁺(CH₃)(ring) $C_5H_{10}N^+$, m/z 84

For 5-aminoquinoline, m/z 144 is the molecular ion, $C_9H_8N_2^+$

(d)

	A	B	C	D	E	F
1	Least-Squares Spreadsheet					
2		x =	y =			
3	Highlight cells B16:C18	µg/L	area ratio			
4	Type "= LINEST(C4:C13,	12	0.056			
5	B4:B13,TRUE,TRUE)	12	0.059			
6	For PC, press	51	0.402			
7	CTRL+SHIFT+ENTER	51	0.391			
8	For Mac, press	102	0.684			
9	COMMAND+RETURN	102	0.669			
10		157	1.011			
11		157	1.063			
12		205	1.278			
13		205	1.355			
14						
15		LINEST output:				
16	m	0.00640	0.02216	b		
17	s_m	0.00019	0.02343	s_b		
18	R^2	0.99333	0.04086	s_y		
19						
20	n =	10	B20 = COUNT(B4:B13)			
21	Mean y =	0.697	B21 = AVERAGE(C4:C13)			
22	$\Sigma(x_i - \text{mean } x)^2$ =	48554.40	B22 = DEVSQ(B4:B13)			
23						
24	Measured y =	0.52	Input			
25	k = Number of replicate measurements of y =	2	Input			
26	Derived x =	77.8	B26 = (B24-$C$16)/$B$16			
27	s_x =	5.0				
28	B27=($C$18/$B$16)*SQRT((1/B25)+(1/$B$20)+((B24-$B$21)^2)/($B$16^2*$B$22))					
29	Measured y =	1.25				
30	k = Number of replicate measurements of y =	2				
31	Derived x =	191.8				
32	s_x =	5.5				

Least-squares parameters are computed in the block B16:C18. In cell B24 we insert the mean y value (0.52) for 2 replicate unknowns. The number of replicates is entered in cell B25. The derived value of x is computed in cell B26. The uncertainty is computed with Equation 4-19 in cell B27. The first term in Equation 4-16 is $1/k$, where $k = 2$ is the number of replicate unknown values. The process is repeated in cells B29:B32 for the second data set.

Answers for the unknowns:

    female nonsmoker: 78 ± 5 µg/L

    nonsmoking girl with smoking parents: 192 ± 6 µg/L

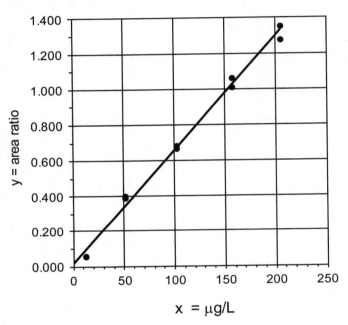

22-28.   (i) b   (ii) a   (iii) c

Electron ionization would be expected to give a molecular ion at m/z 608 plus an isotopic peak with $33 \times 1.1\% = 36\%$ intensity at m/z 609. There would be many fragments at lower mass. Spectrum b fits this description.

Electrospray only gives ions that were already in solution. In a positive ion spectrum, MH^+ at m/z 609 is reasonable to observe. If m/z 609 is MH^+, then there ought to be a peak with ~36% as much intensity at m/z 610. There are not likely to be many more peaks. This description fits spectrum a.

If the base peak from electrospray (m/z 609) goes through a collision cell, it creates smaller fragments. There will be no m/z 610 because the ^{13}C isotopic peak was rejected by the first mass separator. This description fits spectrum c.

22-29. Nitrite: $[^{14}NO_2^-] = [^{15}NO_2^-](R - R_{blank}) = [80.0\ \mu M](0.062 - 0.040) = 1.8\ \mu M$

Nitrate: $[^{14}NO_3^-] = [^{15}NO_3^-](R - R_{blank}) = [800.0\ \mu M](0.538 - 0.058) = 384\ \mu M$

CHAPTER 23
CHROMATOGRAPHIC METHODS AND CAPILLARY ELECTROPHORESIS

23-1. (a) At pH 12, the mixture contains RCO_2^-, RNH_2, Na^+, and OH^-. All of these species pass directly through the cation-exchange column (which is in the Na^+ form). There is no retention of any component.

(b) At pH 3, the mixture contains RCO_2H, RNH_3^+, H^+, and Cl^-. RNH_3^+ cation will be retained by the column and the other species will be eluted.

23-2. One way is to extensively wash with NaOH a column containing a weighed amount of resin to load all of the sites with OH^-. After a thorough washing with water to remove excess NaOH, the column can be eluted with a large quantity of aqueous NaCl to displace OH^-. The eluate is then titrated with standard HCl to determine the moles of displaced OH^-.

23-3. Vanadyl sulfate solution contains the cations VO^{2+} and H^+ with a total concentration of positive charge equal to $2[VO^{2+}] + 2[H_2SO_4]$. When passed through the cation exchange column loaded with H^+, the moles of H^+ eluted are equal to the moles of positive charge applied to the column.

mmol OH^- required to react with eluted H^+ = (13.03 mL)(0.022 74 M NaOH) = 0.296 3 mmol OH^-, which must equal the total cation charge (= $2[VO^{2+}]$ + $2[H_2SO_4]$) in the 5.00-mL aliquot. Fifty milliliters contain 10 times as much charge = 2.963 mmol of cation charge.

From spectrophotometry, we know that
mmol VO^{2+} = (50.0 mL)(0.024 3 M) = 1.21_5 mmol = 2.43_0 mmol of charge

The H_2SO_4 must therefore be $(2.963 - 2.43_0)/2 = 0.26_7$ mmol.

1.21_5 mmol $VOSO_4$ = 0.198 g $VOSO_4$ in 0.244 7-g sample = 80.9 wt%

0.26_7 mmol H_2SO_4 = 0.026_2 g H_2SO_4 in 0.244 7-g sample = $10._7$ wt%

H_2O (by difference) = $8._4$ wt%

23-4. (a) As the pH is lowered, the protein becomes protonated, so the magnitude of the negative charge decreases. The protein becomes less strongly retained.

(b) As NaCl in the eluent is increased, the protein will be displaced from the gel by Cl^- ions.

23-5. The pK_a values are NH_4^+ (9.24), $CH_3NH_3^+$ (10.63), $(CH_3)_2NH_2^+$ (10.77), and $(CH_3)_3NH^+$ (9.80). As the pH is raised from 7, the order in which the cations lose a proton to become neutral species is $NH_3 < (CH_3)_3N < CH_3NH_2 < (CH_3)_2NH$. This order is the anticipated order of elution, because the neutral molecule is not retained by the cation-exchange column. We should not be surprised if the elution order were different because molecular size and hydrogen bonding could affect the selectivity coefficients.

23-6. (a) The suppressor removes eluent ions so that only analyte ions pass through the conductivity detector. If eluent ions were not removed, their conductivity would be greater than that of analyte ions, so analyte would be hard to measure.

(b) The anode generates H^+ that diffuses through the cation exchange membrane and reacts with OH^- from the eluent. The eluent now has excess positive charge from K^+. In the cathode compartment, one OH^- is consumed for each H^+ made in the anode compartment. K^+ from eluent diffuses through the cation exchange membrane into the cathode compartment to replace OH^- that is consumed. The net result is that K^+ from eluent is removed and OH^- from eluent is converted to H_2O.

23-7. In the 250.0 mL solution,
$[IO_3^-] = A/\varepsilon b = (0.521 - 0.049)/[(900 \text{ M}^{-1} \text{ cm}^{-1})(1.000 \text{ cm})] = 5.244 \times 10^{-4}$ M
moles IO_3^- in 250.0 mL = (0.250 L)(5.244 × 10^{-4} M) = 0.131 1 mmol
This much IO_3^- was produced from reaction of 1.000 mL of 1,2-ethanediol solution.

Therefore, 0.131 1 mmol 1,2-ethanediol (= 8.137 mg) must have been consumed.

Weight percent = $100 \cdot \dfrac{(0.008\ 137 \text{ g/mL})(10.0 \text{ mL})}{0.213\ 9 \text{ g}}$ = 38.0 wt%

23-8. (a) $[H^+]$ = 131 μM, so pH = $-\log(131 \times 10^{-6}$ M$)$ = 3.88

(b) Anion charge = $-[Cl^-] - [NO_3^-] - 2[SO_4^{2-}]$ = -773 μM
Cation charge = $[H^+]+[Na^+]+[NH_4^+]+[K^+]+2[Ca^{2+}]+2[Mg^{2+}]$ = $+780$ μM

The two charges agree within 0.9%, so the analysis appears to be complete and accurate. If a major species had been overlooked, the charges would not likely be so well balanced.

(c) Formula masses: Cl^-, 35.45; NO_3^-, 62.00; SO_4^{2-}, 96.06; H^+, 1.01; Na^+, 22.99; NH_4^+, 18.04; K^+, 39.10; Ca^{2+}, 40.08; Mg^{2+}, 24.31

Example: mass of $Cl^- = (101 \times 10^{-6}\ M)(35.45\ g/mol) = 0.003\ 58\ g/L$

Total mass of dissolved ions $= 0.053\ 2\ g/L = 0.053\ 2\ mg/mL$

23-9. At pH 2 (0.01 M HCl), TCA is more dissociated to the carboxylate anion than is DCA, which is more dissociated than MCA.

$$Cl_3CCO_2H\ =\ Cl_3CCO_2^-\ +\ H^+$$

The more negative the average charge of the compound, the more it is excluded from the ion-exchange resin and the more rapidly it is eluted.

23-10. Calibration line: $\log(\text{molecular mass}) = -0.614\ 3\ t_r + 12.666$.

For $t_r = 13.00$ min, $\log(\text{molecular mass}) = 4.680$

Molecular mass $= 10^{4.680} = 4.8 \times 10^4$.

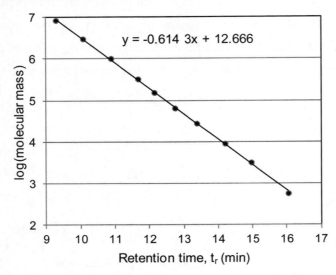

23-11. (a) The total column volume is $\pi r^2 \times \text{length} = \pi(0.39\ cm)^2(30\ cm) = 14.3\ mL$. Totally excluded molecules do not enter the pores and are eluted in the solvent volume (the void volume) outside the particles. Void volume = 40% of 14.3 mL = 5.7 mL.

(b) The smallest molecules that completely penetrate pores will be eluted in a volume that is the sum of the volumes between particles and within pores = 80% of 14.3 mL = 11.5 mL.

(c) Solutes are probably adsorbed on the stationary phase. Otherwise they would all be eluted between 5.7 and 11.5 mL. In addition to size exclusion, this column is carrying out adsorption chromatography.

23-12. (a) Step 1:

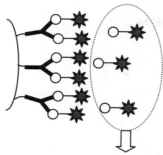
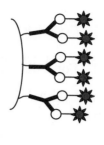

Prior to step 1, silica particles on column contain bound antibodies

At $t = 0$, fluorescence-labeled phenytoin is applied to column to saturate antibodies

Excess fluorescence-labeled phenytoin is washed off, giving signal peak near $t = 2$ min

Step 2:

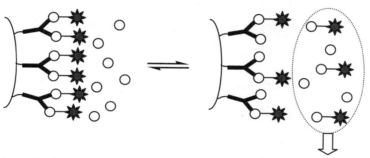

At $t = 6$ min, inject sample containing free phenytoin

Free phenytoin displaces some fluorescence-labeled phenytoin, which is eluted near $t = 8$ min

(b) The first peak near $t = 2$ min is unbound fluorescence-labeled phenytoin that was in excess of the binding capacity of antibodies on the column. The second peak near $t = 8$ min is fluorescence-labeled phenytoin displaced from antibodies by free phenytoin in the unknown serum sample.

(c) Total volume of stationary phase = $\pi r^2 \ell$, where $r = 1.05$ mm is the column radius and $\ell = 0.94$ mm is the length of stationary phase

Volume = $\pi(1.05 \text{ mm})^2 (0.94 \text{ mm}) = 3.26 \text{ mm}^3$

Multiply by $(1 \text{ cm}^3)/(1\,000 \text{ mm}^3)$ to convert volume to 3.26×10^{-3} cm^3
= 0.003 26 mL. But liquid occupies ~50% of total volume ≈ 0.001 63 mL.

Residence time for serum on column ≈ (0.001 63 mL)/(1.2 mL/min)
= 0.001 4 min = 0.08 s.

(d) First compare the standard deviations with the F test:

$$F_{\text{calculated}} = \frac{s_1^2}{s_2^2} = \frac{(0.44)^2}{(0.14)^2} = 9.8_8$$

For 2 degrees of freedom in the numerator and denominator, $F_{\text{table}} = 19.0$. $F_{\text{calculated}} < F_{\text{table}}$, so the standard deviations are not significantly different.

$$s_{\text{pooled}} = \sqrt{\frac{s_1^2(n_1-1) + s_2^2(n_2-1)}{n_1 + n_2 - 2}}$$

$$= \sqrt{\frac{0.14^2(3-1) + 0.44^2(3-1)}{3 + 3 - 2}} = 0.33 \ \mu M$$

$$t = \frac{|\bar{x}_1 - \bar{x}_2|}{s_{\text{pooled}}} \sqrt{\frac{n_1 n_2}{n_1 + n_2}} = \frac{|5.99 - 6.11|}{0.33} \sqrt{\frac{(3)(3)}{3+3}} = 0.44$$

For $3 + 3 - 2 = 4$ degrees of freedom, the critical value of t at 95% confidence is 2.776. Because $t_{\text{calculated}} < t_{\text{table}}$, the difference is not significant.

23-13. (a) Cations before neutrals before anions

(b) At a pH of 4, silanol groups on the capillary will be mostly protonated, so electroosmotic flow will be low. Electroosmotic flow will be from the detector to the injector. Anions will migrate against the electroosmotic flow, from the injector to the detector. Cations will migrate with the electroosmotic flow toward the injector. Cations in the analyte will never reach the detector. (But do not fear! Electroneutrality is maintained by cations in the background electrolyte.)

(c) Electroosmotic flow is driven from the negative injector to the positive detector by anions near the positively charged wall. Superimposed on the electroosmotic flow, anions migrate toward the detector and cations migrate toward the injector. The order of elution is anions before neutrals before cations.

23-14. In the standard mixture, the ratio of peak heights is $\text{height}_{\text{nitrate}}/\text{height}_{\text{iodate}} = 0.92$. Designating nitrate as unknown X and iodate as standard S, we can find the response factor by writing the internal standard equation in the form

$$\frac{A_X}{[X]} = F\left(\frac{A_S}{[S]}\right) \Rightarrow F = \left(\frac{[S]}{[X]}\right)\left(\frac{A_X}{A_S}\right) = \left(\frac{10 \text{ ppm}}{15 \text{ ppm}}\right)\left(\frac{0.92}{1}\right) = 0.61_3$$

in which we have substituted peak heights for peak areas (A).

For diluted aquarium water, height$_{nitrate}$/height$_{iodate}$ = 0.26 and we can write

$$\frac{A_X}{[X]} = F\left(\frac{A_S}{[S]}\right) \Rightarrow [X] = \left(\frac{1}{F}\right)\left(\frac{A_X}{A_S}\right)[S] = \left(\frac{1}{0.61_3}\right)\left(\frac{0.26}{1}\right)[10 \text{ ppm}] = 4.2 \text{ ppm}$$

We just calculated that [NO$_3^-$] = 4.2 ppm in the diluted aquarium water.

The concentration in the aquarium is 100 times greater = 420 ppm or 420 mg NO$_3^-$/L = 6.8 mM.

Here is another way to look at the same calculation without writing out the response factor equation. In the standard solution, the ratio of peak heights is height$_{nitrate}$/height$_{iodate}$ = 0.92. In the unknown, the ratio of peak heights is 0.26. Because [IO$_4^-$] has the same concentration in both solutions, the [NO$_3^-$]$_{unknown}$/[NO$_3^-$]$_{standard}$ = 0.26/0.92 = 0.28. But [NO$_3^-$]$_{standard}$ = 15 ppm, so [NO$_3^-$]$_{unknown}$ = (0.28)(15 ppm) = 4.2 ppm.

23-15. (a) Electroosmosis is the bulk flow of fluid in a capillary caused by migration of cations in the diffuse part of the double layer toward the cathode.

(b) At pH 9, there is a high concentration of negative Si—O$^-$ groups on the wall and a correspondingly high concentration of cations in the diffuse part of the double layer to create electroosmotic flow. At pH 3, much of the negative charge on the wall is reduced by the reaction Si—O$^-$ + H$^+$ = Si—OH. With less negative charge on the wall, there is less positive charge in the double layer to drive electroosmotic flow.

(c) A capillary coated with Si—OC$_{18}$H$_{37}$ does not readily ionize to make Si—O$^-$. The capillary wall has little negative charge at any pH.

23-16. An ionic component of eluent with the same sign of charge as the analyte is chosen to give a large, constant detector response. Analyte gives lower detector response. When analyte moves past the detector, the concentration of background electrolyte decreases (to maintain electroneutrality), so the detector signal decreases.

23-17. In the absence of micelles, neutral molecules are all swept through the capillary at the electroosmotic velocity. Negatively charged micelles swim upstream with some electrophoretic velocity, so they take longer than neutral molecules to reach the detector. A neutral molecule spends some time free in solution and some time dissolved in the micelles. Therefore, the net velocity of the neutral molecule is reduced from the electroosmotic velocity. Different neutral molecules have different solubilities in the micelles and spend a different fraction of the time in the micelles and free in solution. Therefore, each type of neutral molecule has its own net migration speed.

Micellar electrokinetic chromatography is a form of chromatography because the micelles behave as a "stationary" phase in the capillary. The concentration of micelles is uniform throughout the capillary. Analyte partitions between the mobile phase and the micelles as the analyte travels through the capillary.

23-18. van Deemter equation:

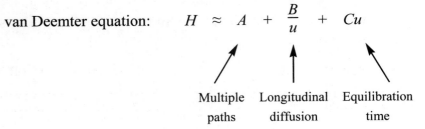

(a) In capillary zone electrophoresis under ideal conditions, longitudinal diffusion is the only source of zone broadening. Therefore, the van Deemter equation should have the form $H \approx B/u$, where u is migration velocity and B is a constant. The graph is shown below.

(b) In micellar electrokinetic capillary chromatography, bands broaden by longitudinal diffusion and from the finite rate of mass transfer (slow equilibration time) between free solution and the micelles. Therefore, the van Deemter equation should have the form $H \approx B/u + Cu$, where u is migration velocity and B and C are constants:

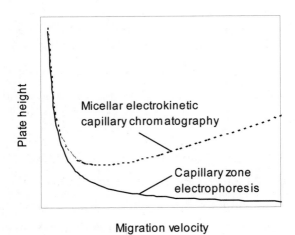

23-19. (a) I measured $t = 40.1$ min and $w_{1/2} = 0.75$ min for $^{35}\text{Cl}^-$.

$$N = \frac{5.55 t^2}{w_{1/2}^2} = \frac{5.55(40.1 \text{ min})^2}{(0.75 \text{ min})^2} = 1.6 \times 10^4 \text{ plates}$$

(b) H = plate height = $40 \text{ cm}/1.6 \times 10^4$ plates = 25 μm/plate

(c) Cl^- has no absorbance at 254 nm, whereas background electrolyte (chromate) absorbs at this wavelength.

23-20. In the absence of micelles, the expected order of elution is cations before neutrals before anions: thiamine < (niacinamide + riboflavin) < niacin. Because thiamine is eluted last, it must be most soluble in the micelles.

23-21.

	A	B	C	D	E	F
1	Molecular mass by SDS/capillary gel electrophoresis					
2						
3		Molecular		Migration	Relative migration	
4	Protein	mass (MM)	log(MM)	time (min)	time (t_{rel})	$1/t_{rel}$
5	Marker dye	low		13.17		
6	a-Lactalbumin	14200	4.152	16.46	1.250	0.8001
7	Carbonic anhydrase	29000	4.462	18.66	1.417	0.7058
8	Ovalbumin	45000	4.653	20.16	1.531	0.6533
9	Bovine serum albumin	66000	4.820	22.36	1.698	0.5890
10	Phosphorylase B	97000	4.987	23.56	1.789	0.5590
11	b-Galactosidase	116000	5.064	24.97	1.896	0.5274
12	Myosin	205000	5.312	28.25	2.145	0.4662
13	Ferritin light chain	?		17.07	1.296	0.7715
14	Ferritin heavy chain	?		17.97	1.364	0.7329

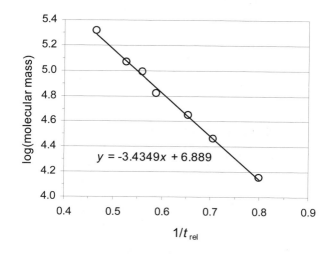

To find the molecular mass of the light chain, we insert its value of $1/t_{rel}$ into the equation of the straight line and solve for log(molecular mass):

$$\log(\text{molecular mass}) = -3.4349(0.7715) + 6.889 = 4.239$$

$$\text{molecular mass} = 10^{4.239} = 17\,300 \text{ Da}$$

Similarly, for the heavy chain we find molecular mass = 23 500 Da

23-22. (a) For a BrO_3^- concentration of 0.2 µg/L, the relative standard deviation of 14.4% corresponds to $(0.144)(0.2 \text{ µg/L}) = 0.02_{88}$ µg/L.

Limit of detection = $(3)(0.02_{88} \text{ µg/L}) = 0.08_6$ µg/L

Limit of quantitation = $(10)(0.02_{88} \text{ µg/L}) = 0.2_9$ µg/L

Results for the other concentrations:

Bromate concentration (µg/L)	Relative standard deviation (%)	Concentration standard deviation (µg/L)	Detection limit (µg/L)	Quantitation limit (µg/L)
0.2	14.4	0.028 8	0.08$_6$	0.2$_9$
0.5	6.8	0.034 0	0.10$_2$	0.3$_4$
1.0	3.2	0.032 0	0.09$_6$	0.3$_2$
2.0	1.9	0.038 0	0.11$_4$	0.3$_8$
		mean:	0.10	0.3$_3$

(b) A bromate concentration of 0.10 µg/L corresponds to $(0.10 \text{ µg/L})/(127.90 \text{ g/mol}) = 0.78_2$ nM. The Br_3^- concentration will be 3 times greater (= 2.3_5 nM) because 1 mol of BrO_3^- makes 3 mol Br_3^-. The absorbance in a 0.600-cm cell will be

$$A = \varepsilon bc = (40\,900 \text{ M}^{-1} \text{ cm}^{-1})(2.3_5 \text{ nM})(0.600 \text{ cm}) = 0.000\,058$$

23-23. (a) If everything else is the same, the greater the charge on an ion, the faster it will migrate in the electric field. For two molecules with similar size and shape, we expect that the greater their difference in charge, the greater their difference in migration time.

(b)

	A	B	C	D	E
1	Charge difference between two weak acids				
2					
3	pKa (acid 1) =	pH	alpha (acid 1)	alpha (acid 2)	difference
4	2.97	3	0.517	0.028	0.489
5	pKa (acid 2) =	3.5	0.772	0.084	0.689
6	4.54	3.7	0.843	0.126	0.717
7		3.7550	0.859	0.141	0.718
8		3.9	0.895	0.186	0.708
9		4	0.915	0.224	0.691
10		4.5	0.971	0.477	0.494
11		5	0.991	0.743	0.248
12	C4 = 10^-$A$4/(10^-B4+10^-$A$4)				
13	D4 = 10^-$A$6/(10^-B4+10^-$A$6)				
14	E4 = C4-D4				

The fraction of an acid HA in the charged form A^-, which we designate α_{A^-}, was given by Equation 12-19:

$$\alpha_{A^-} = \frac{[A^-]}{F} = \frac{K_a}{[H^+] + K_a}$$

where F is the formal concentration and K_a is the acid dissociation constant. The spreadsheet computes the fraction of A^- in columns C and D for each of the acids and the difference between the two fractions in column E. The difference is greatest at pH 3.755, midway between the two pK_a values.

23-24. Pentylammonium cation is anchored to the stationary phase because the hydrophobic tail is soluble in the C_{18} phase. The cationic headgroup behaves as an ion exchange site for anions. Cations in the analyte mixture are not retained by the cations fixed to the stationary phase. However, the analyte anions are retained to varying degrees and, therefore, separated. Sulfate is eluted last because it has the most negative charge and is retained most strongly by the anion exchange column.

23-25. (a) Filtration ensures that solid particles larger than 0.45 µm are not included in the analysis. Only soluble or colloidal species are analyzed.

(b) Anion exchange resin retains SO_4^{2-} from the large volume of rainwater. When SO_4^{2-} from 30 L of rainwater is displaced from the column with 300 mL of 3 M NaCl, SO_4^{2-} is concentrated by a factor of 100.

(c) Addition of $BaCl_2$ precipitates $BaSO_4$, which is collected at the end of the experiment. Radioactive ^{35}S in the precipitate is counted with a scintillation counter.

(d) Na_2SO_4 acts as a *carrier* to ensure that there is enough mass of product to handle at the end of the analysis. If carrier had not been added to the initial sample, there would have been too little solid to isolate at the end of the experiment.

23-26. Perhaps Mg^{2+} and Ca^{2+} cations neutralize the negative charge on the wall of the capillary by binding tightly to $-Si-O^-$ groups. If the wall is uncharged, there will be no electroosmotic flow. Apparently, EDTA binds Mg^{2+} and Ca^{2+} tightly enough to prevent them from binding to the wall.

$$\text{Mg} \diagdown \text{O} \diagup \text{O} \diagdown -Si-O-Si-$$